James Pillans

First Steps in the Physical and Classical Geography of the Ancient World

Thirteenth Edition

James Pillans

First Steps in the Physical and Classical Geography of the Ancient World
Thirteenth Edition

ISBN/EAN: 9783744728386

Printed in Europe, USA, Canada, Australia, Japan

Cover: Foto ©berggeist007 / pixelio.de

More available books at **www.hansebooks.com**

FIRST STEPS

IN THE

PHYSICAL AND CLASSICAL GEOGRAPHY

OF THE

ANCIENT WORLD

BY

JAMES PILLANS, LL.D.

LATE EMERITUS PROFESSOR OF HUMANITY IN THE UNIVERSITY OF EDINBURGH.

THIRTEENTH EDITION.

Revised, Enlarged, and Illustrated with Maps,

BY

THOMAS FAWCETT, B.A.

QUEEN'S COLLEGE, OXFORD, HEAD MASTER OF BLENCOWE GRAMMAR SCHOOL.

EDINBURGH:
ADAM AND CHARLES BLACK.

MDCCCLXXXII.

EDINBURGH:
PRINTED BY H. AND J. PILLANS,
12 THISTLE STREET.

PREFACE TO THE THIRTEENTH EDITION.

In this Edition, while mainly adhering to the principles laid down by Professor Pillans in the Preface to his excellent work, I have added several places of great importance in Ancient History,—thus increasing the work by several pages of printed matter. Among others I may mention the Allia, Cremera, Munda, Thapsus, Regillus, Arginusæ, Chæronea, Ægospotamos, which are second to none in the annals of Greece and Rome. New plans of Rome and Athens, with explanatory text, appear now for the first time, and a map of the neighbourhood of Rome to illustrate the early Roman History. Names of importance are printed in thick type in order to catch the eye. Having found by experience with my own boys that the map prepared by Professor Pillans though excellent in execution, was much too difficult—nay impossible—for them to decipher, mainly from its being so overcrowded, especially in such important countries as Italy and Greece, and that boys wasted much valuable time in finding the places, I have added separate skeleton maps for each country, which, I think, will be a great improvement. By continuing the map of Asia eastward

I have included Arbela, Cunaxa, Carrhæ, and Babylon. The localities recorded in the text have numbers placed after them, and these will be found in the map. By this means schoolboys before their lessons and private students generally, may examine themselves as to the position of the places in each country, without the treacherous aid of the name before their eyes. As the work now stands, it will be found to contain all that is required in an elementary course of Ancient Geography. A skeleton wall-map is published by Messrs W. & A. K. Johnston, Edinburgh, price 10s., which will be found of great service for class teaching.

BLENCOWE GRAMMAR SCHOOL,
10th *May* 1882.

PREFACE TO THE SECOND EDITION.

IN the Introduction to a larger work, on the same subject as this and intended for more advanced students in Schools and Colleges, I have laid down the principles on which, as it appears to me, all geographical knowledge ought to be acquired by the learner, and communicated by the teacher: and, in the body of that work, those principles are applied to classify and illustrate the details of whatever in this branch of knowledge was deemed important to be taught in a course of classical education.*

Geography, however, may be made eminently attractive to young minds at an earlier stage than that which is contemplated in the larger work; and, when applied from the very outset to the illustration of the classics read, it contributes not a little to give interest, facility, and permanence to the instruction conveyed. It is with this end in view that I have put together these 'First Steps.' They are confined within a com-

---

* "Elements of Physical and Classical Geography:" to which are prefixed, 1. A Sketch of the Planetary System of which the Earth we inhabit is a part 2. A short account of the Structure of our Globe and the Changes it has undergone; and 3. A Series of Astronomical Tables. To which is added, an Anthology of Passages from the Classics, illustrative of the Text. Edinburgh and London, 1854.

paratively small space on the surface of the globe. But, when we consider the multiplicity of studies to which, in these days, the attention of our educated youth is directed, it will probably be found that this little volume contains the ground-work of as much instruction in Ancient Geography as it is possible to overtake in School or College, within the limited time at the disposal of either.

In compiling and *compressing* this syllabus,—for the small number of the pages is but an imperfect measure of the labour bestowed on the task,—I have been guided by *three* principles, which may be thus stated :—

I. When the main object is to throw light upon the classics and give an interest to classical studies, it is neither necessary nor desirable, in an ordinary course of school or college training, to go beyond the countries, some portion of which is on the shore either of the Mediterranean itself, or of one of those seas which are in truth parts of it, though called by distinct names,—the Adriatic, the Aegean, the Propontis, and the Euxine.

In accordance with this view, I here invite the student to accompany me—staff in hand, as it were, and *right* shoulder to the sea—from one of the Pillars of Hercules, at the southern extremity of Spain, to the other, at the north-western extremity of Africa. In making this tour, the moment we set foot on the soil of a new country, we quit the coast for a time, and explore the interior in every direction; noting, as we go along, those physical characters and localities which are most fertile in classical associations, and to which interesting allusions are most frequently made by the writers and especially the poets of antiquity, or by the most admired poets of our own island.

Then, resuming our journey along the coast, from the point where we left it to explore the interior, we add to our previous enumeration of localities the Towns, Sea-ports, Capes, and River-mouths that may be worthy of notice; and, before we pass into another country to repeat there the same processes, it may be found convenient, in helping us to a more accurate knowledge of the places, to make ourselves acquainted with the most important of the ancient subdivisions of the territory which we have been examining.

As to countries beyond the range which this tour, so conducted, will make us acquainted with, they were either imperfectly known to the ancients and therefore seldom alluded to, or their geography will be most advantageously studied, either somewhat later, or not till the student comes to read the authors whose writings refer to them.

II. The second principle which has guided me in this selection is, that, in impressing on the memory the localities and relative positions of every place enumerated, the physical aspects and external conformation of the country are to be kept in view and referred to, in preference to the conventional distribution of the surface into civil districts and provinces. The former are sensible realities and permanent characters: the latter are arbitrary, ideal, and fluctuating. In carrying out this principle, the first thing to be done is, to set before the learner's eye and so imprint on his mind, a lively image or representation of the country in question, composed of the line of Coast,—the groups and ranges of Mountains,—the main Rivers, with their principal tributaries,—and the River-Basins through which they all flow. And this is best done by coloured chalks on a black board.

It is then, and not till then, that we proceed to fill up this outline with the details of the picture; to trace the main rivers from their sources downwards, marking, as we descend, the cities and towns of note on their banks. And thus both town and river are fixed in the memory, by making each suggestive of the other. Other towns are next attached in the same way to the principal tributaries; and the remainder, if any, we refer to the line of coast or the river-basin in which they are situated. By this process we assign 'a local habitation and a name' to every thing that is memorable, and give it a hold upon the imagination and a facility in being recalled, which no arrangement according to civil divisions can possibly attain

III. In the last place, I have made it a principle, in selecting the things worthy of note, that no town or locality shall be inserted, to the mention of which is not appended some fact, circumstance, or peculiarity, which may not only give it a chance of being treasured up in the memory, but be likely to awaken in enquiring minds a desire to know more about the place and its history. These little touches indeed are intended as texts for the teacher to prelect and enlarge upon; and if this be done dexterously and felicitously, the impressions made will be both pleasing and permanent, and will furnish land-marks for the learner to steer by, should he afterwards wish to acquaint himself more fully with the country they belong to.

In every stage of geographical instruction, where the student receives his knowledge from others and does not work it out for himself, all names of places which are inculcated without such appendage as I now speak of, or with nothing more than assigning them to the shire or county they are in, are little better than useless rubbish. They are indeed worse than useless; for these *voces et praeterea nihil* serve only to encumber the memory and disgust the learner with a subject which, if properly treated, is full of interest to a young mind. The circumstance of towns being on the same river, or included in the same river basin, is much more likely to be remembered than the fact of their belonging to the same civil division. But even that circumstance I have not considered as sufficient, unless coupled with some physical, historical, or commercial memorandum, and of a kind as striking and interesting as possible.

In selecting the circumstances and associations which are most likely to rivet the prominent features and localities of a country in the memory of youth, I have borrowed largely from the beautiful fancies and fables of the Heathen Mythology. It may be well however to apprize the reader, that I am no convert to a practice which has been recently introduced in works of great merit and high authority; that, I mean, of substituting for the Latin names of the ancient divinities, which have been familiar to us all as household words from our infancy, the corresponding Greek terms and printing them in the characters of the English alphabet; while the Roman designations are either discarded altogether, or degraded to a secondary place and imprisoned within brackets. Our old friends Jupiter and Juno are scarcely recognizable in their new titles and costumes as Zeus and Hera : the God of Fire limps into his forge in Mt Aetna under the familiar name of Vulcan or the holiday appellation of Mulciber, and limps out again with the portentous title of Hephaestus : Neptune, it seems, must resign his Trident, and it is to be feared his Planet too, to Poseidon : Mercury, should he indulge his old propensity to thieving may escape from justice under the *aliàs* of Hermes ; and the Great Globe itself, the common nursing-mother of us all, who rejoiced in the double honours of Tellus and Terra, is curtailed of her fair proportions, and appears under the humiliating monosyllabic misnomer of Ge. There is, in short, a complete change of ministry in the councils of Olympus. I confess myself a firm adherent of the old Administration, and live in the hope of seeing it once more in office; but, in the meantime, it may not be amiss to give the reader a tabular view of the two Cabinets :—

PREFACE. xi

Old.		New.
Jupiter superseded by		Zeus.
Juno	...	Hera.
Minerva	...	Athēna.
Vulcan	...	Hephaestus.
Neptune	...	Poseidon.
Pluto	...	Hades.
Mars	...	Ares.
Ceres	...	Demēter.
Venus	...	Aphrodite.
Cupid	...	Eros.
Diana	...	Artĕmis.
Bacchus	...	Dionȳsus.
Latona	...	Leto.
Tellus—Terra	...	Ge v. Gaea.

There is another innovation, to which, though not of quite so recent a date, I feel equally disinclined to give in:—the practice, to wit, of printing the ancient names of places in the ordinary type, and the modern names in the Italic character. This is the reverse of D'Anville's rule, who in all his printed works uses the Italic for the ancient nomenclature, and the common or Roman letter for the modern. And this is a rule which has reason on its side no less than authority. For surely it is a more natural and convenient arrangement, to have the modern names of localities in the same type as the modern text, and to reserve for the ancient names that character which has itself an antiquated look, and which, though now rarely used, was the prevailing method at the era of the invention of printing. And it did so prevail, because it bore the closest resemblance to, and was in truth an imitation of, the much prized manuscripts of the classics. What motives induced the editors to reverse this method in their excellent 'Compendium of Geography' now in use at Eton College,* I have not been able to learn. They are not stated in the Preface; nor can I well imagine what they were, or why the example thus set has been so generally followed in later publications. For myself, having adopted D'Anville's rule forty years ago, and abided by it down to the recent edition of the 'Elements,' I am unwilling to abandon a practice which can plead the sanction and the example of the prince of modern geographers.

* First published by Arrowsmith in 1831.

COLLEGE OF EDINBURGH,
June 1855.

FIRST STEPS

IN THE

PHYSICAL AND CLASSICAL GEOGRAPHY

OF THE

ANCIENT WORLD.

I.

HISPANIA (Graecé et *Poetice* **Iberia**, *hodie*, SPAIN and PORTUGAL),

WAS the name given by the Romans to a peninsula of quadrangular shape, in length and in breadth about 600 miles, which occupies the S.W. extremity of Europe, and is wholly contained within the lines of 36° and 44° N. Latitude, and of $3\frac{1}{2}$° E. and $9\frac{1}{2}$° W. Longitude.

PHYSICAL CHARACTERS OF THE PENINSULA.

An elevated ridge of Mountain and Table-land extends from N. to S., forming the water-shed of the country, and giving origin to all the great rivers; of which some find their way to the Mediterranean, and some to the Atlantic. To this crest, or back-bone as it were of the country, are attached, on the side facing the west, ranges of mountains and high ground running in a S.W. direction and nearly parallel to each other, which enclose, on two sides, the Basins or tracts of country, through which the rivers and their tributaries flow.

HISPANIA

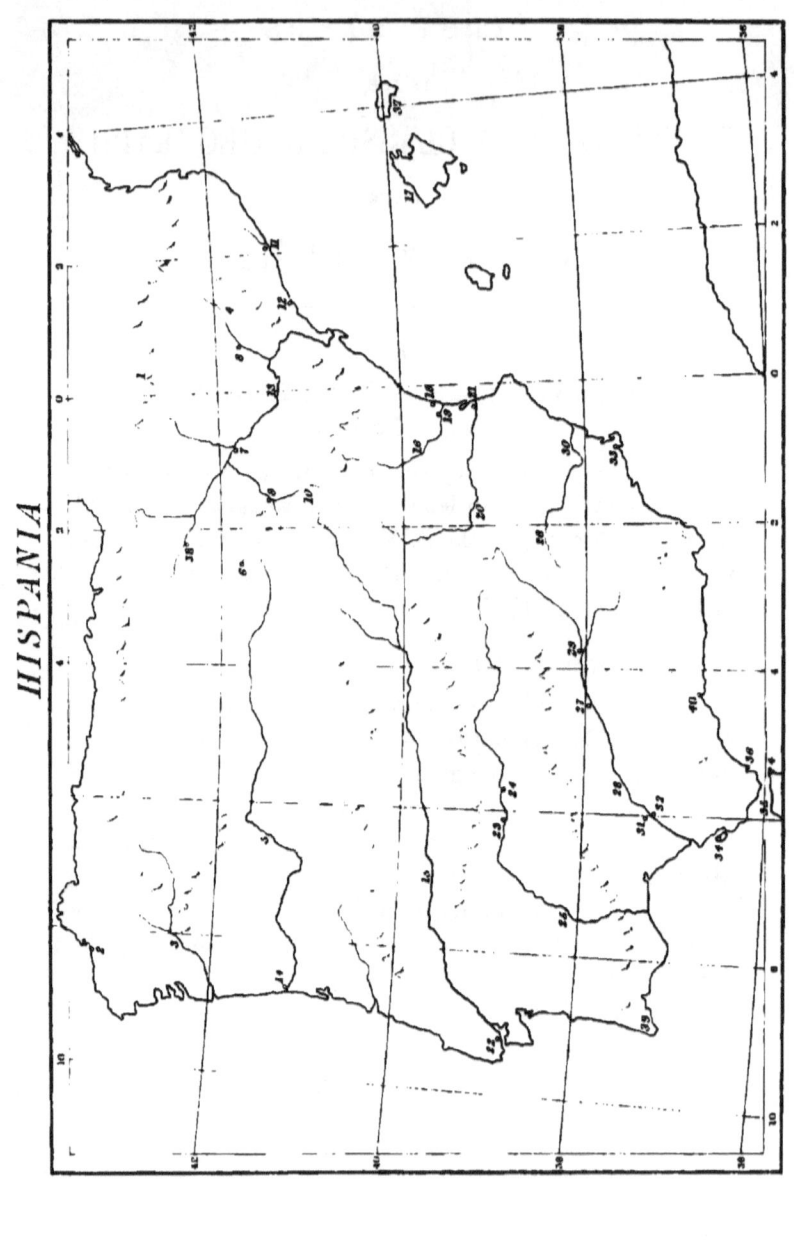

The main Rivers on the W. side of the central ridge, and falling into the Atlantic, are *four* in number : 1. **Durius** (5), the Duéro (in Spanish), Douro (in Portuguese); whose vast basin, bounded by the Cantabrian and Asturian mountains on one side and by those of Castille on the other, includes the less considerable valley of **Minius** (3), the Minho;—2. **Tagus** (15), the Tajo or Tagus ;—3. **Anas** (25), the Guadiana;—and 4. **Baetis** (28), the Guadalquivir (Wadalkiveer).

The main rivers on the E. side of the water-shed and falling into the Mediterranean, are also *four*, but excepting the last, of much shorter course : **Tader** (26), the Segura ; **Sucro** (20), the Xucar ; **Turia** (16), the Guadalaviar ; and **Iberus** (13), the Ebro : And the basins of these rivers are enclosed in like manner by lateral ranges of hills which start off, like spinal processes, from the Eastern side of the central ridge.

In tracing the rivers enumerated, *secundo flumine*, from fountain-head to mouth or *embouchure*, we find in succession the following towns and localities :—

1. On the **Durius**, near the source, **Numantia** (6), (*Hispaniae decus*), which sustained a fourteen years' siege, and was taken at last by Scipio Africanus Minor, B.C. 133.* At the mouth was **Calle** (14) or *Portus Calensis*, whence the kingdom of PORTUGAL derives its name. *Calle* is now Oporto, and from this comes the word 'Port,' as applied to wine shipped from that harbour.

2. On a tributary of the **Tagus**, now called Manzanares, stands MADRID, the modern capital of Spain ; and not far from the embouchure, on the N. side, was **Olysipo** (22), now LISBON, the capital of Portugal.

* Ille Numantinâ traxit ab urbe notam.—*Ovid. Fast.* i. 596.

3. On the **Anas**, **Metellinum** (24) (Medellin), probably so called from Caecilius Metellus, the founder; and **Emerita Augusta** (23), Merida, a town built by Augustus to reward his veterans (*emeriti*).

4. On the **Baetis**, near the source, was **Castulo** (29), (Cazlona), of which Himilce, the wife of Hannibal, was a native: the country round was *Saltus Castulonensis*, part of the Table-land of Sierra Morena, the scene of the fabulous adventures of Don Quixote; farther down the river, **Corduba** (27), (Cordova), birthplace of LUCAN and the two SENECAS; **Italica** (31), birth-place of the Emperor Trajan, and some think, of Hadrian also, and the poet Silius Italicus; **Hispalis** (32), SEVILLE, which ranks as the second city in modern Spain.

5. **Tader**, the Segura, is the furthest south of the *four* main rivers which fall into the Mediterranean. After passing the modern city of Murcia, it flows through the **Campus Spartarius** (30), a plain so called from its abounding in *spartum* (esparto), a reed much used by the ancients for the cordage of ships and various economical purposes.*

6. **Sucro** (20), the Xucar, had at its mouth a city (21), of the same name (πολις ομωνυμος, *Strab.*), where a mutiny once broke out in the Roman army which was quelled by Scipio Africanus Major.†

7. At the embouchure of **Turia** (Guadalaviar), was **Valentia** (19), a Roman colony, now the capital of VALENCIA, a Spanish province unequalled in natural advantages. It is called by the natives La Huerta

---

* See Plin. Nat. Hist. xix. 2; Liv. xxii. 20; Hom. Il. B. ii. 135.

† The story is finely told by Livy, B. xxviii., ch. 24, &c.

(*hortus*), and wants nothing but good government and enterprise to make it the garden of Europe.

8. On the **Iberus** we meet with **Calagurris** (38), (Calahorra), memorable for the dreadful sufferings of Sertorius' army, when besieged by Pompey and Metullus, B.C. 75. Half way down stood **Salduba** (7), afterwards **Caesar-Augusta,** now ZARAGOZA, made illustrious in the Peninsular war by its successful resistance to the French invaders in 1808-9.

The broad Basin of the IBERUS, lying between the Pyrenees and the Central Ridge, is watered, from the heights of both, by numerous tributary streams, the most remarkable of which are, on the N. side, the **Sicoris** (4), (Segre), on which stood **Ilerda** (8), (Lerida), where Caesar defeated Pompey's generals Afranius and Petreius, B.C. 46 ; and on the S. side, **Salo** (10), (Xalon) on which stood **Bilbilis** (9), the native town of the poet Martial.

AFTER thus following the courses of Rivers, if we next take for our guide the line of Coast, we shall fall in with towns which have been indebted for their importance and notoriety, in ancient or modern times, to the convenience of harbourage, and the facility of access and resort to commercial and colonising foreigners.

In this tour of the Coast, starting round from C. Ortegal, the N.W. angle of the peninsula, and moving South, we find the town and harbour of **Corunna** (2), (one of the *Portus Artabrorum*), called by British traders the Groyne ; where Sir John Moore fell in the moment of victory (Jan. 1809). 'Corunna' is thought to be a corruption of **Columna,** from an ancient tower still standing, said to have been built by Hercules. Half way between the mouth of the *Tagus* and the *Anas* we come upon the **Sacrum Promontorium** (39), where the sun was supposed to plunge into the sea (*Juv.* xiv. 280). At the S.W. angle of the peninsula, between the mouth of the *Baetis* and the Strait of Gibraltar (**Fretum Herculeum**) (35), stood the very ancient town of *Gadir*, founded and so named by the Phoenicians. Among the Romans it was called **Gades** (34), and considered as the extreme point of the earth Westward ('solisque cubilia *Gades*'), in like manner as the Ganges was reckoned the farthest limit Eastward. When Juvenal says,—

> Omnibus in terris quae sunt a *Gadibus* usque
> Auroram et *Gangen*.—x. 1.

he means to express the entire *length* of the earth. *Gadir* is the modern town and harbour of Cadiz.

Within the Strait is **Calpe** (36), (the Rock of Gibraltar), which the poets feigned to be one of the Pillars erected by Hercules as his *meta laborum*, when he had reached the western *terminus* of the habitable globe.

Proceeding now along the shore of the Mediterranean, we come to **Munda** (40), memorable (1.) for the defeat of Carthaginians by Scipio, B.C. 216 ; (2.) for Caesar's victory over the younger Pompey, B.C. 45. Then to **Nova Carthago** (33), (Carthagena), the capital of the Carthaginian possessions in Spain, till it was taken by Scipio Africanus Major, B.C. 210.* A little way north of Valencia was **Saguntum** (18), ('urbs illa, fide et acrumnis inclita'), the storming of which was Hannibal's first act of aggression in the Second Punic war. Out of its ruins was built the modern town of Murviedro, *i.e. muri veteres*.

* See Livy's interesting account of the capture, B. 26, c. 42-6.

Between the mouth of the Ebro and the Pyrenees were **Tarraco** (12), (Tarragona) chief city of the Roman province *Tarraconensis*, and **Barcino** (11), (Barcelona), said to have been built by Hamilcar Barcas, father of Hannibal.

Off the coast of Valencia, is the group of **Baleares Insulae** (17), famed for furnishing corps of slingers to the Roman armies. In *minor* we have **Portus Magonis** (37), (now Port Mahon), so called from Hannibal's brother Mago.

Continuing our journey Northward from Barcelona, we cross the Eastern extremity of **Montes Pyrenaei** (1), (Pyrenees), and find ourselves in Gallia Transalpina.

GALLIA TRANSPALPINA.

THIS is a portion of the earth's surface lying wholly within the lines of N. Lat. 42° and 52° and of $4\frac{1}{2}$° W. and $8\frac{1}{2}$° E. Long. The term comprehends not only the country of the **Helvetii** (0), and other Alpine tribes lying to the *left* of the Rhine, but the whole territory on the left side of that river from its source to its mouth, so that Gaul extended about 700 miles in length as well as in breadth. The Mountain ranges of Gaul which rise high enough to deserve the name, are the following: 1. **Gebenna** (47), the Cevennes, stretching N.N.E. from the Pyrenees; 2. an extinct volcanic group in Auvergne (**Arverni**) (38), the highest points of which are the Cantal, Mont Dor, and Puy-de-Dome; 3. **Vogesus** (18), the Vôges, running parallel with the Rhine from Bâle to Coblentz; 3. **Jura** (31), which formed the boundary between the *Helvetii* and the **Sequani** (29): And 5. all that portion of the Alps which lies to the W. and S. of the Upper Rhine, and which sends the water produced in its summits and slopes either into that river, or into the Rhone.

The vast superficial extent of Ancient Gaul (very considerably larger than Modern France) may be regarded as composed of six large BASINS, (*i.e.* tracts of land penetrated throughout their whole length by a main river and its tributaries); and these basins are separated from each other, either by the mountains above enumerated, or by high grounds (called *dos* in French from the Latin *dorsum*), which serve equally well the purposes of water-shed. The Basins are, those of the **Garonne** (43),

GALLIA

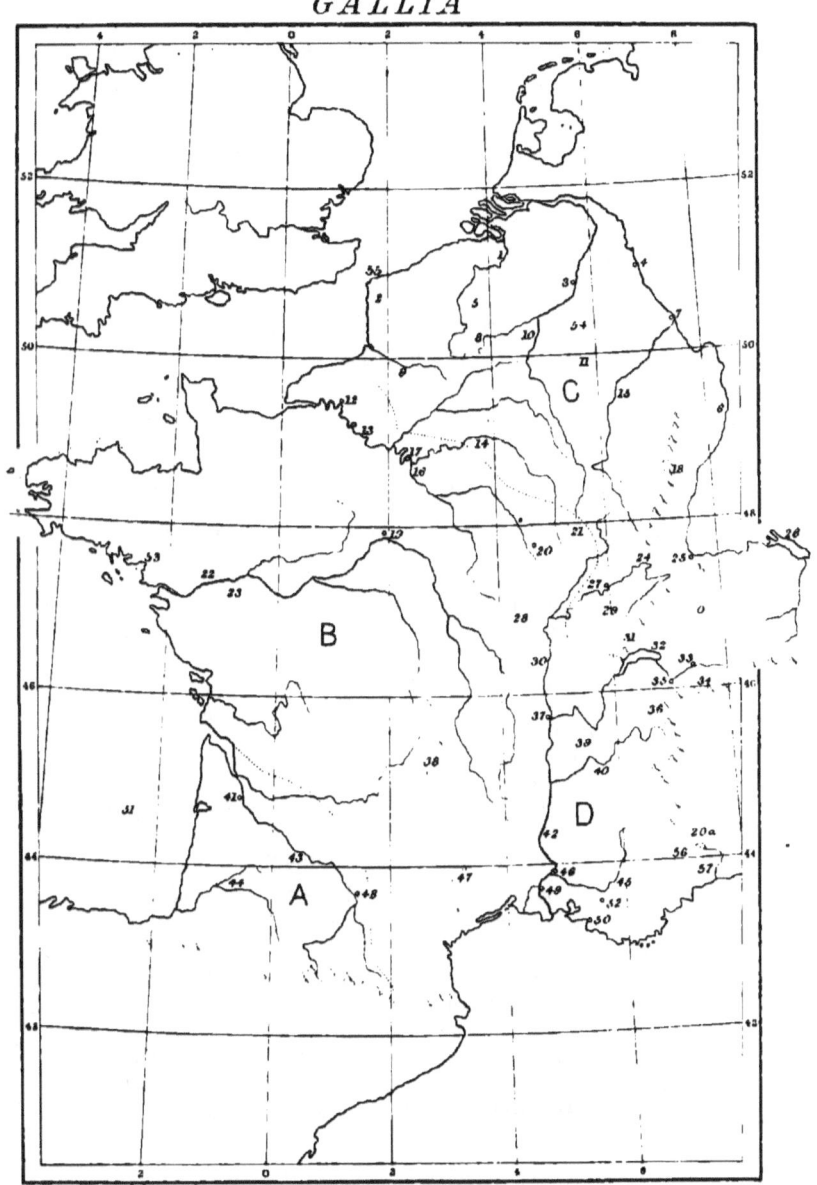

the **Loire** (23), the **Seine** (13), the **Meuse** (10), the **Rhine** (6), and the **Rhone** (42). The basins of the rivers will be found to account for the whole superficial contents of *Gallia Transalpina*, except the district watered by the Somme (**Samara**) (9), and the Scheld (**Scaldis**) (1), which are little more than 'rivières de côte.' We have then beginning from the S.W.,

1. The BASIN of **Garumna** (43), the Garonne, a river which rises in the Pyrenees, and flows N.W. into the Bay of Biscay (**Sinus Cantabricus**) (51). This basin is bounded by the Pyrenees, the Cevennes, the Mountains of Auvergne, and the *dos* or high ground on the N. Within these limits it includes the minor basin of **Aturus** (44), (the Adour). In descending the *Garumna* we find **Tolosa** (48), Toulouse;* and farther down, on the left bank of

---

* Toulouse has been a seat of learning both in ancient and modern times. Martial gives it the epithet *Palladia*, as well on that account as for its temple of Minerva. It was the birth-place and residence of the famous jurist Cujacius, and of Maynard, distinguished as a man of letters and a courtier, but so ill rewarded, that he retired at last to his humble home and inscribed over the door of the cabinet he died in, the following lines :—

> " Las d'espérer, et de me plaindre
> Des Muses, des Grands, et du Sort,
> C'est ici que j'attends la mort,
> Sans la désirer, ni la craindre."

TOULOUSE will be for ever memorable as the scene of the final action and crowning victory of that series of Peninsular campaigns, which, taken in connection with the battle of Assaye that preceded them, and the battle of Waterloo that came after, have fixed the name of WELLINGTON as the first of all commanders of armies. And if we also take into account the tenor and purpose of his whole life, the magnitude and importance of the transactions in which he was engaged, and, above all, the uniform subordination of self-interest to a sense of duty,—for in him the love of money, the love

the river, stood **Burdegala** (41), the modern city of Bordeaux, well known for its commerce and its claret (vin de Bordeaux). Lower down, the Garonne receives the Dordogne, and widening into an estuary is called La Gironde.

2. The BASIN of **Liger** (23), the Loire. This river rises in the Cevennes, and flows, Northward first and then Westward, into the Atlantic, which it reaches after a course of 500 miles. Among the Towns on its banks most worthy of mention was **Genabum** (19), which owes its modern name of Orleans to the people **Aureliani** (19), whose capital it was; or, as some think, to the Emperor Aurelian. This town has been made famous in modern times by the story of the Maid of Orleans, and by its giving title to the first prince of the blood in the Old Monarchy of France. Near the embouchure of the Loire dwelt the tribe **Namnetes** (22), who have given name to the modern city of Nantz, (in French, Nantes). The revocation of the edict of Henri IV., and the consequent influx of so many French protestants into Britain, led to Nantz being adopted and spelt as an English word. A little to the north dwelt the **Veneti** (53). The help

of power, and even that 'last infirmity of noble minds,' the love of fame, yielding a willing obedience to the love of country and the observance of right,—we shall not only place him above the vulgar herd of conquerors and founders of dynasties, but regard him as the greatest among the men of all ages who have been called on to act a conspicuous part, at once in the civil and military affairs of nations. Such we may presume will be the judgment of posterity, so long as it shall be deemed a nobler exercise of talent and more worthy of renown, to save one's country than to enslave it. In this class of characters, Washington, though a man scarcely equal in talent, stands alone on the same high level of moral greatness with Wellington.

sent by the Britons to this tribe gave occasion to Caesar's invasion, B.C. 55.

3. The BASIN of **Sequana** (13), the SEINE. This river rises in the table-land, called the Plateau de Langres, of the Gallic tribe **Lingones** (21). Near the source was **Alesia** (20), a town taken by Caesar after a long siege, which he describes minutely in the 7th B. of his Commentaries. A little way below its junction with **Matrona** (14), (Marne), it encloses an islet called **Lutetia Parisiorum** (17), now in the very heart of PARIS the capital of France. On the river, between Paris and the sea, is Rouen (**Rotomagus**), (12), b. pl. of the great Corneille.

4. Passing over the Somme and the 'lazy' Scheld, we arrive at the BASIN of **Mosa** (10), (in Dutch, Maas, in French, Meuse), on which, as we descend the river, we come successively upon Liège, Namur, and Maestricht (**Mosae Trajectus**), (3) places of little note in ancient times, but whose names occur often in the history of modern wars.

5. The BASIN of **Rhenus** (6), the RHINE, of which the *left* side only is Gallic. The Rhine rises in the central Alps, and its early course is enclosed between Alpine ranges, until it expands into **Lacus Brigantinus** (26) or **Venetus**, the Lake of Constance. Thence it flows Westward (forming, at Schaffhausen, the most noted waterfall in Europe) till it reach **Basilia** (25), (Basel or Bâle), and meeting there with an obstruction in the high ground between Jura and the Vôges, it turns to the North. In the subsequent part of its course it passes the walls, *first* of **Confluentes** (7), corrupted into the modern Coblentz, at the *confluence* of the Rhine and **Mosula** (15) v. *Mosella* (Mosel or Moselle); and *then*, of **Colonia Agrippina** (4),

Cöln or Cologne, with its famed Cathedral. Upon one filament of that network of ditches, canals, and inlets of the sea, in which the Meuse and Rhine lose themselves in the latter part of their course, stands the modern city of Rotterdam, the b. pl. of Erasmus, to whom his fellow-citizens have erected a bronze statute on one of the bridges.

6. The BASIN of **Rhodanus** (42), the RHONE. This river, rising in **Mons Adula** (34), (near the Pass of St Gothard), makes its way between the two loftiest ranges of the Alps through the Vallais, where it passes the city of the **Seduni** (33), now Sitten or Sion, and **Octodurus** (35), Martigny. Then, forcing its way through the gorge of St Maurice, it expands into **Lacus Lemanus** (32), the Lake of GENEVA, resumes its river form at the town of that name, and after emerging from a subterranean channel a quarter of a mile long called *la perte du Rhone*, proceeds Westward, till, meeting with the obstruction of the Cevennes, it turns abruptly to the South. At this angle it is joined from the N. by **Arar** (30), (Saone), which Caesar describes as flowing *incredibili lenitate*. At the point of junction stood **Lugdunum** (37), which gave name to one of the Augustai divisions of Gaul, *Lugdunensis*. This city, under the modern name of Lyon, is famed, among other things, for its silk manufactures, and has long ranked next to Paris in importance and population. From Lyon the Rhone continues its rapid course directly S., passing various towns, among which may be mentioned **Avenio** (46), (Avignon), at the junction of **Druentia** (45), (Durance), and **Arelate** (49), (Arles), where the river separates into two branches, enclosing a Delta of rich land called Camargue, (perhaps

a corruption of *Caii Marii ager*). About ten leagues E. of this stood **Massilia** (50), (Marseilles), said to have been founded at a very remote period by a colony from *Phocaea*, a town on the coast of Asia Minor. As a dependency of Rome, *Massilia* rose to great prosperity and refinement. Tacitus mentions it as the place of Agricola's education, and calls it "locus Graeca comitate et provinciali parsimoniâ mistus ac bene compositus." Above *Massilia* was **Aquae Sextiae** (52), (Aix), where Marius defeated the Teutoni, B.C. 102.

Ancient Divisions of GALLIA.

At the time of Caesar's invasion, (B.C. 58), there was already in Gaul a **Provincia Romana** (D), lying between the Cevennes and the Alps. The rest of Gaul he describes as divided into three parts, according as it was inhabited by **Aquitani** (A), in the South, **Belgae** (C), in the north, and **Celtae** (B), between the two. But the truth is that **Gallia Comata**, as all beyond the Roman Province was then called, was occupied by numerous independent tribes or peoples, generally hostile to each other. Some of these have been already named, such as the **Lingones** (21), and **Parisii** (16), in the basin of the Seine, the **Aureliani** (19), and **Namnetes** (22), in that of the Loire, and the **Seduni** (33), in the Valais. A few shall be now added as occurring most frequently in Caesar's narrative of his campaigns in Gaul: and the locality of each tribe named will be referred to the river-basin in which it dwelt.

The **Aedui** (28), ('clarissimi Celtarum,') occupied the territory between the Loire and the Saone; the **Sequani** (29), the upper part of the basin of the Saone, and the whole of that of its feeder the **Dubis** (24), Doubs, on which river was their chief city **Vesontio** (27), (Besançon). The **Allobroges** (39), dwelt between the Rhone and its left-hand tributary **Isara** (40), the Isère. The **Treveri** (11), v. *Treviri* occupied the space between the Meuse and the Rhine, and the lower basin of the Moselle. Their chief city was that now called from the name of the tribe, —in German, Trier,—in French, Trèves. To the north of this tribe was the celebrated wood **Arduenna** (54), described, Caes. v. 3, known to modern readers as the seat of the "wild boar of Ardennes." To the west, occupying the basin of **Sabis** (8), (Sambre), a tributary of the Meuse and the upper course of the Scheld, dwelt the **Nervii** (5), a gallant people of German extraction, who made head against Caesar in a great battle, and, but for his own prowess and presence of mind, would have gained the victory*. Hence one of the proudest recollections of his life was

'That day he overcame the Nervii.'†

The battle was fought on the banks of the same river (the Sambre) along which Napoleon marched in his way to the field of Waterloo. Still further West, on the Strait of Dover and Calais, lived the people commemorated by Virgil in the line

"Extremique hominum **Morini** (2), Rhenusque bicornis."
Aen. viii. 727.

Near the coast is **Portus Itius** (55), whence Caesar sailed on his expedition to Britain.

* Bell. Gall. ii. cap. 15—28.
† Shaksp. Jul. Caes. Act iii. sc. 2.

16 DIFFERENT TRIBES.

Having finished our survey of Gaul, we return to the point at the East end of the Pyrenees whence we commenced it, and proceed along the shore of *mare nostrum*, as the Romans called the Mediterranean, till we arrive at the **Alpes Maritimae** (56), and the little river **Varus** (57), the Var; on crossing which we find ourselves in ITALY.

ITALIA.

(*Graecé* et *Poeticé*, Hesperia, Oenotria, Ausonia, Saturnia *Tellus*) including **Gallia** *Cisalpina*, and **Magna Graecia**.

Italia, in the widest acceptation of the word, (in which however it was not used till the days of Imperial Rome), comprehended the whole of that territory which is fenced off to the N.W. from the rest of Europe by the mountain barrier of the Alps, and is surrounded on all other sides by the sea. It extends 700 miles in length, and is of various breadth, lying between the parallels of 38° and 47° N. Lat. and the lines of 7° and 19° E. Long.

ITALY, when contemplated under its physical aspect, is composed of two portions nearly equal in extent, but widely different in natural character.

The one is the Peninsula of *Italia Propria*, surrounded by the waters of the Mediterranean and Adriatic on all sides, except to the N.W. where an imaginary line overland connects the little streams of **Macra** (87), and **Rubicon** (88), and forms the isthmus.

The other portion is mainly the great basin of **Padus** (15), called also by the poets *Eridănus*,—the Po.

Between these two portions of Italian territory there is a striking contrast. In the Northern division, throughout its whole length, we find a river flowing in the lowest level between the Alpine and Apennine heights which form its boundaries. In the Southern or peninsular portion, the reverse is the case. The central line of

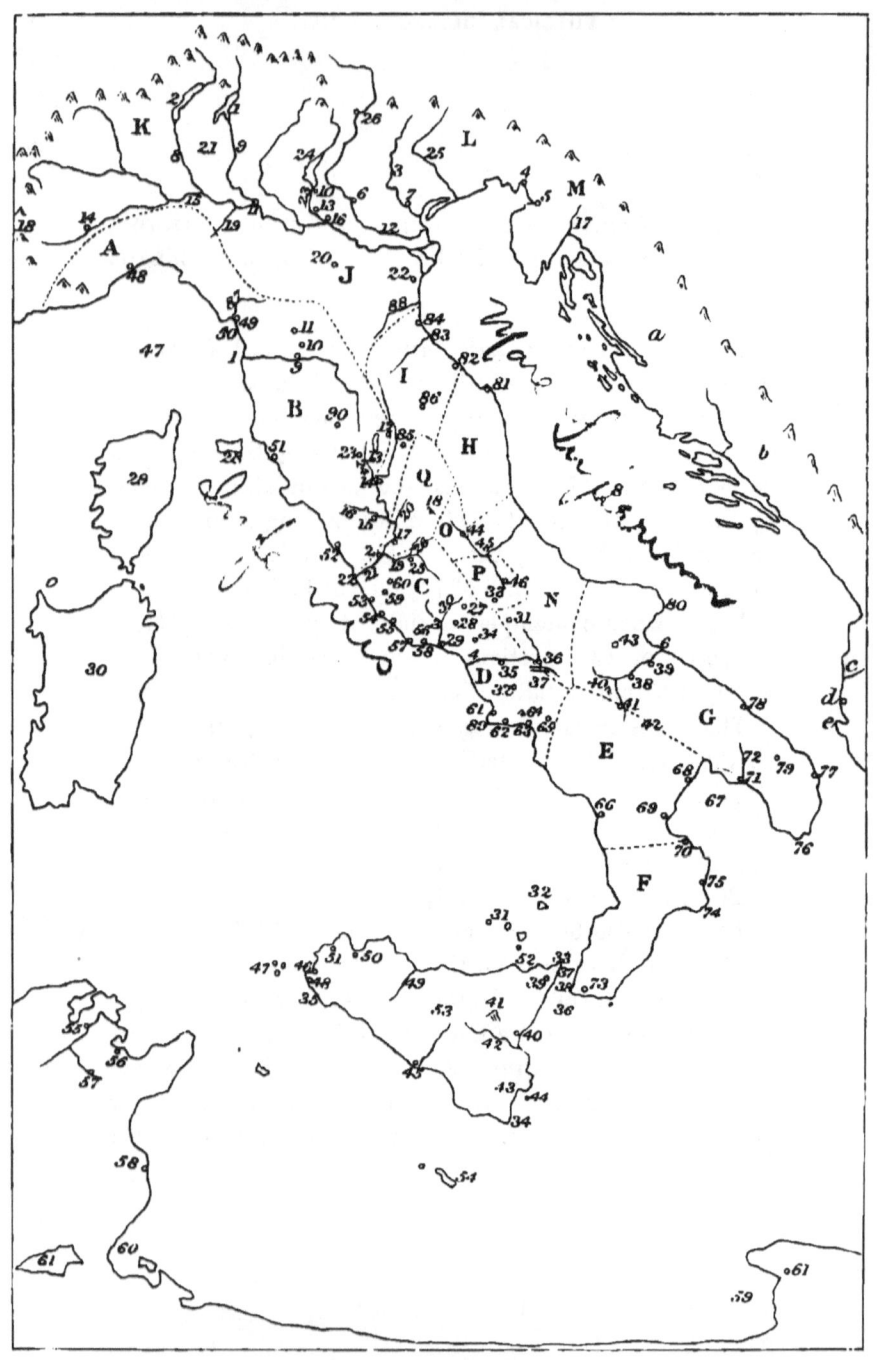

the Peninsula is not the lowest but the most elevated part, it being in fact the crest of the lofty and continuous chain of the Apennines; while the boundary line on the two sides is the lowest of all levels—the sea. The one region is penetrated by a single river, swollen by the contributions of innumerable streams from the opposite sides of the great basin, all of which find their way to the Po, the great receptacle which absorbs them and pours their united waters into the Adriatic. The Peninsula, on the other hand, has abundance of streams, but they are all, even the Tiber, of comparatively short course, having each its own little basin and lateral feeders, and falling directly and independently into the sea.

To begin with **Italia Propria**.—The only rivers of considerable length are, the **Arnus** (1), **Tiberis** (2), **Liris** (3), **Vulturnus** (4), which flow into the **Mare Inferum** (5), vel *Etruscum*, and the **Aufidus** (6), and **Aternus** (7), which flow into **Mare Superum** (8), vel *Adriaticum*...

1. In the BASIN of **Arnus** (Valdarno) and Vallombrosa)* was **Florentia** (9), now Florence, capital of Tuscany, and near it **Faesulae** (10), where the 'Tuscan artist' Galileo made his observations, and **Pistoria** (11), where Catiline was defeated and slain.

2. In the BASIN of **Tiberis** (2), the Tiber, (in Italian, Tevĕre), were:—

(1.) On the river itself, **Perusia** (12), near **Lacus Trasimenus** (13), (now the lake of Perugia), where Hannibal routed the Roman army a third time under

* Thick as autumnal leaves that strew the brooks
In Vallombrosa, where the Etrurian shades
High over-arch'd embower.—*Par. Lost*, i. 302. See also 287.

Flaminius; considerably lower down was **Fescenninum** (14), where a coarse kind of drama was performed; and below that the well known city of **Veii** (15), on the **Cremera** (16), where the *Fabii* were cut off; **Fidenae** (17), near **Mons Sacer** (18), between the *Tiber* and **Anio** (19); a little North was the river **Allia** (20), where Brennus defeated the Romans, 390 B.C.: **Roma** (21)—*Princeps Urbium, Urbs Septicollis :*—and, at the mouth, **Ostia** (22), the Port of Rome.

(2.) On the *right* side of the basin, **Clusium** (23), the city of Porsenna, on the tributary **Clanis** (24); and on the *left* side, **Tibur** (25), (Tivoli) on the *Anio* (Teverone). At Tibur Maecenas had a villa where Horace was a frequent guest, his Sabine farm being at no great distance, on the **Digentia** (26), (Licenza) a feeder of the *Anio*.

3. In the BASIN of **Liris** (Garigliano) were, on the *left* side **Arpinum** (27), birth-place of Marius and Cicero, the famous *Duo Arpinates;* **Aquinum** (28), birth-place of Juvenal; at the mouth, near the Marshes where Marius took refuge, **Minturnae** (29). On the same side, the *Liris* was joined by **Fibrenus** (30), on whose banks and in the little island at the junction which belonged to Cicero, was held the Dialogue de Legibus, (see *De Legg.* lib. ii. 1-3).

4. In the BASIN of the **Vulturnus**, on the *left* side of the river, stood the following towns :—**Allifae** (31), famed for its pottery—(*Allifāna,* sc. pocula, seem to have been remarkable for their size) :—**Capua** (32), chief city of the *Campāni,* and the rival of Rome itself (hence called 'altera Roma') till towards the close of the Second Punic war, when, having taken part with Hannibal, it fell with his falling fortunes: On the *right* side was **Venafrum** (33),

famed for its olives, and **Cales**(-*ium*) (34) for its vines; (*Venafranum*, sc. oleum, and *Calēnum*, sc. vinum, denoted oil and wine of the first quality). **Beneventum** (36), a town of *Samnium*, on the *Via Appia*, stood at the point of junction of *Sabātus* and *Calor*, whose united stream falls into the *Vulturnus*.

On that river itself stood **Casilinum** (35), (on the site of the modern Capua), which gained credit with the Romans by its long and obstinate resistance to Hannibal, (*Liv.* B. xxii. ch. 15.) Between *Beneventum* and *Capua*, lay **Furcae Caudinae** (37), a defile where a Roman army was hemmed in by the Samnites under Pontius, and forced to pass under the yoke, (*Liv.* B. ix. ch. 1-9.)

5. In the BASIN of the **Aufidus**, not far from the *right* bank of the river, were **Canusium** (38) and **Cannae** (39): Near the latter was gained the last and greatest victory of Hannibal, B.C. 216; and to *Canusium* the poor remains of the Roman army retreated after the disastrous battle. Higher up the valley, to the south of Mt. **Vultur** (40), was **Venusia** (41), b.-pl. of Horace, on the debateable land between **Apulia** and **Lucania**; hence Horace speaks of himself as 'Lucanus an Appulus anceps.' Here also was the **Fons Bandusiae** (42); not, as usually supposed, in his Sabine farm. To the north, **Asculum** (43), the scene of Pyrrhus' great victory over the Romans, B.C. 297.

6. In the BASIN of the **Aternus**, on the river itself, were **Amiternum** (44), b.-pl. of Sallust the historian, and **Corfinium** (45), the rallying point of the League against Rome in the Social War. At some distance south from the bend of the river, stood **Sulmo** (46), (Sulmona) a town of the *Peligni*, b.-pl. of Ovid.

To the geographical position of other towns and localities not connected with these six Rivers a clue will be found, if we follow the line of coast, having special reference at the same time to the following sub-divisions or provinces of **Italia Antiqua**.

These Provinces are either Maritime or Inland. Of the Maritime, six bordered on the Mediterranean, and six on the Adriatic. The former were **Liguria** (A), **Etruria** (B), **Latium** (C), **Campania** (D), **Lucania** (E), and the **Bruttii** (F); those on the Adriatic were **Apulia** (G), (including *Japygia* and *Daunia*), **Picenum** (H), **Umbria** (I), **Gallia** (J) **Cispadana, Gallia Transpadana,** (K), and **Venetia** (L), including the peninsula of **Istria** (M). The Inland Provinces were **Samnium** (N), and the Highland districts of the **Marsi** (O), **Peligni** (P), and **Sabini** (Q).

Let us first travel along the coast of the *maritime* provinces in the above order :—

1. On the coast of **Liguria**, which was the name applied to the stripe of land between the Apennines and the sea, extending from the Var to the Macra,—we find, at the head of the Bay called **Sinus Ligusticus** (47), **Genua** (48), a town more famous in history under its modernised form of GENOA:

2. Crossing the **Macra** (87), we enter **Etruria**, and arrive at the Town of **Luna** (49), and its harbour **Portus Lunensis** (50), (Gulf of Spezzia), that which, Lucan affirms—'non est spatiosior alter, Innumeras cepisse rates, et claudere pontum ;' and not far off are the quarries of Carrara, which still furnish statuary marble to Europe. **Telamon** (51), where a great battle was fought between the Romans and the Gauls, B.C. 225. As

we approach the mouth of the Tiber, we come upon the ancient **Agylla** (52), afterwards called *Caere*, a town rewarded with the honorary freedom of the City, for its fidelity to Rome at the time of the Gallic invasion :

3. On the coast of **Latium**, the first town we meet with is **Laurentum** (53), the City of King Latinus, next **Lavinium** (54), and then **Antium** (55), the Capital of the *Volsci*. It was over the *Antiātes* that the Romans gained their first victory at sea : in memory of which they fixed the beaks (*rostra*) of the ships they had captured, in front of the tribune from which the orators harangued the people. *Antium* was famed in Horace's time for a temple of Fortune. Beyond this were **Paludes Pomptinae** (56), the Pontine Marshes, a tract of country where the *malaria*, so prevalent in many parts of Italy, is peculiarly noxious. Next come the Town Promontory and Harbour of **Cajeta** (57), (Gaëta), which took the name, Virgil tells us, from the nurse of Aeneas. Near it was Cicero's **Formianum** (58), where he was murdered by order of Mark Antony. Here commences and is continued into *Campania*, the district in which the choicest wines of the ancients were produced,—the *Formiani Colles*, the *Mons Massĭcus*, the *Ager Falernus, Caecŭbus, Calēnus, Setīnus*. Inland were several small lakes, the chief of which were **L. Albanus** (59), and **L. Regillus** (60), the latter being the scene of a great battle between the Romans and the Latins, B.C. 496, where the gods fought, Troy-fashion, in the likeness of men.

4. On the coast of **Campania**, were 1. **Cumae** (61), which Virgil makes the first landing-place of Aeneas in Italy and the abode of the Sybil who conducted him to the shade of his father Anchises, in the abodes of the dead:

—2. **Baiae** (62), a favourite watering-place : *—3. **Parthenope** (63), afterwards *Neapŏlis* (Napŏli, NAPLES), one of those Greek Colonies which were so numerous along this southern shore of the peninsula that it got the name of **Magna Graecia**. At a little distance across the bay on which Naples stands is **Vesuvius** (64), a volcano, of which the first eruption upon record took place A.D. 79, when the elder Pliny lost his life; and it has continued ever since to be the only active volcano in continental Europe. At the base of Vesuvius, and overwhelmed by its erruption, were the buried cities *Herculaneum* and *Pompeii*, discovered and partially disinterred within the last and present centuries. At no great distance, inland, was **Nola** (65), at the siege of which Hannibal for the first time received a check, (*Liv.* 23. 16.) It was at *Nola* that Augustus died :

5. On the coast of **Lucania** was **Paestum** (66), famed for its roses and its ruined temples, 'Biferique rosaria Paesti.' On that part of the Lucanian coast which is in the **Sinus Tarentinus** (67), were **Metapontum** (68), the residence for a time of Pythagoras, and of Hannibal;— **Heraclea** (69), the place of assembly for the deputies from the states of **Magna Graecia**, where Pyrrhus defeated the Romans, B.C. 281 ; and **Sybaris** (70), proverbial for the luxury and effeminacy of its inhabitants. In the same bay, but beyond the limits of Lucania, was **Tarentum** (71), on the brook **Galesus** (72) : †

6. In **Ager Bruttius**, on the **Fretum Siculum** (36),

* Nullus in orbe sinus *Baiis* praelucet amoenis.
—*Hor. Epist.* i. 1. 83.

† Dulce pellitis ovibus Galesi
Flumen et regnata petam Laconi
Rura Phalanto.—*Hor. Od.* ii. 6.

(Strait of Messina) was a rock and cave under it, fabled to be the residence of the sea-monster **Scylla** (38) ; farther along, in the narrowest part of the Strait, was the Town of **Rhegium** (73), supposed to have received its name from the tradition of Sicily having been there *broken* off from Italy (ἀπὸ τοῦ ῥαγῆναι, *Strabo*).

Near the **prom. Lacinium** (74) was **Croton** (75), where Pythagoras taught his doctrines ; the b.-pl. also of the noted athlete, Milo Crotoniates.

7. On the Adriatic coast of **Apulia**, after doubling Cape **Japygium** (76), (Leuca) we find **Hydruntum** (77), the shortest transit to Greece, but less frequented for that purpose than **Brundusium** (78), which had an excellent harbour, and was the terminus of the *Via Appia*, which was the great high road from Rome to Greece. Brundusium, and **Dyrrhachium** (D), on the opposite coast, were the Dover and Calais of the ancient world. This part of the **Apulian** coast was inhabited by a people called *Calabri*: their Town **Rudiae** (79) was the b.-pl. of the poet Ennius, who is hence called by Cicero *Rudius homo*, and his poetry Horace calls *Calabrae Pierides*. Then comes the projection of the land occupied by **Mt. Garganus** (80), and its oak forests (*quercēta Gargani*, Hor.) Horace compares the uproar in a Roman theatre to a storm among the woods of Garganus :

Garganum mugire putes nemus.—*Epist.* II., i. 202.

8. On the coast of **Picenum**, which was celebrated for its apples ('Picenis cedunt pomis Tiburtia succo,') we fall in with a similar projection of the land, which, from the form it takes, was likened to the human elbow, ἀγκών,

* For an interesting account of *Croton*, and the Temple of Juno Lacinia in its neighbourhood, see Liv. xxiv. 3.

and hence the town built upon it got the name of **Ancon** (81) vel *Ancōna*, which it still retains, ('Dalmaticis obnoxia fluctibus *Ancon.*'—Lucan.)

9. On the coast of **Umbria** were two towns of note, 1. **Sena** (82), to which the epithet *Gallica* was added, as well to denote the fact of its being originally a Gallic settlement, as to distinguish it from **Sena Julia** (90), an inland Town in Etruria; the former is now Sinigaglia, the latter Sienna; and, after crossing the **Metaurus** (83), where Hasdrubal was defeated and slain by Salinator and Nero, B.C. 207, we come to 2. **Ariminum** (84), (Rimini), the storming of which was Caesar's first overt act of civil war, after crossing the **Rubicon** (88). In the south of Umbria, but inland, we have **Interamna** (85), the b.-pl. of Tacitus, the Roman Historian, and near the centre, **Sentinum** (86), where Q. Fabius defeated the Samnites and Gauls, B.C. 295.

We will now proceed with the Northern Division, *i.e.* the vast basin of the Po. During the Republican times, it was no part of Italy, but was known to the Romans as **Gallia** *Cisalpina*. If we trace the **Padus** (15), from its source in **Mons Vesulus** (18) (Monte Viso) to its embouchure, we shall find on the river itself, 1. **Augusta Taurinorum** (14), taken by Hannibal on his descent from the Alps, (now TURIN) (TORINO), capital of the State of Sardinia. To the north, between the Ticinus and Addua are the **Raudii Campi** (21), where Marius defeated the Cimbri, B.C. 100. And 2. **Cremona** (11), whose vicinity to Mantua is lamented by Virgil : * a city noted in modern times for the excellence of the violins manufactured there.

The North side of the Po Basin, from its position in

* Mantua, vae! miserae nimium vicina Cremonae !—*Ecl.* ix. 28.

regard to Rome, was called **Gallia Transpadana** (K), a region watered by numerous Alpine tributaries. The most remarkable are, 1. **Ticinus** (8) (Tesino), issuing from **Lacus Verbanus** (2), (Lago Maggiore), on whose banks Hannibal first defeated the Romans in a skirmish of cavalry, B.C. 218 : 2. The **Addua** (9), (Adda), issuing from Lake **Larius** (1), (Lago di Como) : and 3. **Mincius** * (23), which drains the superfluous waters of Lake **Benacus** (24), (Lago di Garda). It issues from the lake, close to the little peninsula of **Sirmio** (10), the favourite residence of the poet Catullus; and on its way to the Po, invests **Mantua** (13), a city which Silius Italicus calls *musarum domus* as being the b.-pl. of Virgil, though it is believed that the poet was born at **Andes** (16), a neighbouring village.

The South side of the Po Basin, as being that nearest to Rome, was called **Gallia Cispadana** (J). It was watered by many tributary streams, among the rest by the **Trebia** (19), on the banks of which the Romans sustained from Hannibal a second and more severe defeat, B.C. 218. Almost in the centre of the district stood **Mutina** (20), now Modena, where D. Brutus was besieged by M. Antony, but relieved by Hirtius and Pansa, the last of the *five* consuls, B.C. 43.

10, 11. On the coast of **Gallia Cisalpina**, south of the Po, stood **Ravenna** (22), near which Augustus constructed a station for his fleet on the *Mare Superum*, as he did at **Misenum** (89), near Naples, to guard the *Mare Inferum :*

12. **Venetia**, though it has rivers of its own which do

---

* ———— tardis ingens ubi flexibus errat
Mincius, et tenera praetexit arundine ripas.—*Virg. G.* iii. 14.
Smooth-sliding *Mincius*, crowned with vocal reeds.—*Lyc.* i. 86.

not fall into the Po, may yet, with reference to the Alpine boundary, be reckoned part of the great basin of Cisalpine Gaul. Its Rivers of note were :—

1. **Athesis** (12), (in German, Etsch, in French Adige), on which were **Tridentum** (26), Trent, memorable for the Council of Catholic Bishops assembled there immediately after the Reformation, A.D. 1545–63, and **Verona** (6), b.-pl. of Catullus, where there is still a Roman amphitheatre in tolerable preservation ; 2. **Medoacus** (3) minor, on which was **Patavium** (7), (Padua), b.-pl. of Livy ; 3. **Timavus** (25), see Aen. i. 244 ; 4. **Arsia** (17), in the peninsula of **Istria**, is the Eastern Boundary of Italy. The City of VENICE, on the coast N. of the Po, belongs to Modern Geography.

ITALIAN ISLANDS *of Note.*

OFF the coast of Etruria, **Ilva** (28), (Elba), famed of old for the richness of its iron ores, ('Insula inexhaustis Chalybum generosa metallis,') and in recent times, as the temporary place of banishment of Napoleon : S.W. of *Ilva* are 1. **Corsica** (29), the native island of Napoleon ; used occasionally by the Romans as a place of exile; 2. **Sardinia** (30), the *Sardo* of the Greeks, called by them also *Ichnūsa*, from its fancied resemblance to the impress of a human foot (ἴχνος, vestigium).* It was noted for a bitter plant, which, from its distortions of the mouth, gave rise to a 'Sardonic laugh,' Virg. Ecl. 7, 41.

East of *Sardinia*, and near the S. extremity of Italy, lies the group of volcanic islets called **Ins. Aeoliae** (31)

* Humanae in speciem plantae se magna figurat
  Insula . . . dives ager frugum.—*Claud. Bel. Gild.* 507.

v. *Vulcaniae* (Lipari Islands), one only of which, **Strongyle** (32), (Stromboli) is still active.

South of this group, lies **Sicilia**, called also *Trinacria*, v. '*Trinācris*, a positu nomen adepta loci.' The three promontories, (τρία ἄκρα, trina cornua, *Ov.*) at the three corners of the triangular island were N.E. **Pelorus** (33), (C. Faro), S.E. **Pachynus** (34), (Passaro), and W. **Lilybaeum** (35), (Boëo).

In the strait which separates Italy and Sicily **Fretum Siculum** (36), the poets describe a whirlpool called **Charybdis** (37), opposite to **Scylla** (38), on the Italian side. These were the two bugbears of ancient navigators, between which it was thought so difficult to steer, that in avoiding the one it was hardly possible not to fall a prey to the other. Hence came the proverbial use of the *modern* line, 'Incidit in Scyllam qui vult vitare Charybdin.'

On the coast between *Pelorus* and *Pachynus* are 1. the town of **Zancle** (39), originally so named from ζάγκλη, a sickle, which the form of the harbour suggested, afterwards *Messāna*, now Messina; 2. **Catine** (40), Catania, which has suffered repeatedly from the lava of the burning mountain, alike famed in fable and in history, **Mt. Aetna** (41): and 3. after crossing **Simaethus** (42), the river of longest course in the Island, we reach **Syracusae** (43), the renowned Metropolis of *Sicilia*, memorable (1.) for defeat of Athenians by the Lacedaemonians, Thucyd. B. 7, B.C. 413; (2.) for its capture by Marcellus, B.C. 212. In front of the harbour is the Island **Ortygia** (44) v. *Nasos*, and in it the fountain *Arethusa*, of poetical celebrity.

Between *Pachynus* and *Lilybaeum* was **Agrigentum** (45), or, in the Greek form **Acragas**, the second city in ancient Sicily; an early rival of Carthage, and noted for a Temple of Jupiter, of which some gigantic fragments still remain. The ancient name survives in the modern Girgenti.

Between *Lilybaeum* and *Pelorus*, on the northern shore of the Island, the notable localities are, **Eryx** (46), a Town, and Mountain; the latter surmounted by a Temple of Venus (Erycīna). Off shore, **Aegates insulae** (47), where the Romans gained the naval victory which put an end to the First Punic War. **Drepanum** (48), (Trapăni), so called, like Zanclé, from the form of its harbour, (δρεπάνη, meaning a scythe). **Himera** (49), where 300,000 Carthaginians were terribly defeated by Gelo on the very day that Salamis was fought. **Panormus** (50), now Palermo, the modern capital of Sicily. **Segesta** (51), famous in history as having occasioned the Athenian expedition to Sicily, B.C. 415. Considerably to the East, near Pelorus, was **Mylae** (52), (1.) where Agrippa defeated the fleet of Sextus Pompeius, B.C. 36; and (2.) where C. Duilius defeated the Carthaginians, B.C. 260, in honour of which, the Columna Rostrata, bristling with beaks (*rostra*) of Carthaginian ships was set up in the Forum. In the centre of the Islands, was

'that fair field
Of **Enna** (53), where Proserpina gathering flowers,
Herself a fairer flower, by gloomy Dis
Was gathered.'—*Par. Lost*, iv. 268.*

To the south of Sicily was **Melita** (54) or *Malta*, the scene of St Paul's shipwreck.

Resuming now our continental journey from VENICE, along the head of the Adriatic, and passing **Aquileia** (4)

* Cicero describes the place minutely, in Verr. de Signis, c. 48.

and **Tergeste** (5), (Trieste), both Roman colonies on the coast of **Istria**, we cross the **Arsia** (17), and bid adieu to Italy. And now,

> 'ILLYRICI legitur plaga littoris, arva teruntur Dalmatiae.'—*Claudian. De iii. Cons. Honor.* v. 119.

Illyricum consisted mainly of a stripe of land between the Adriatic and a range of mountains branching off from the Eastern Alps and running S.E. and then East, under various names, *Albii Montes, Scardus, Scomius, Pangaeus, Rhodŏpe,* and at last *Haemus,* a name which it retains till it reaches the Euxine. **Liburnia** (A), was the northern and **Dalmatia** the southern province of **Illyricum** (B). Skirting the latter we cross the river **Drilo** (C), and, coming in sight of the 'infames scopulos **Acroceraunia**, (E), find ourselves at last on the soil of that country, in which it may be said, with the least poetical exaggeration, that 'not a mountain rears its head unsung.'

VICINIA ROMANA

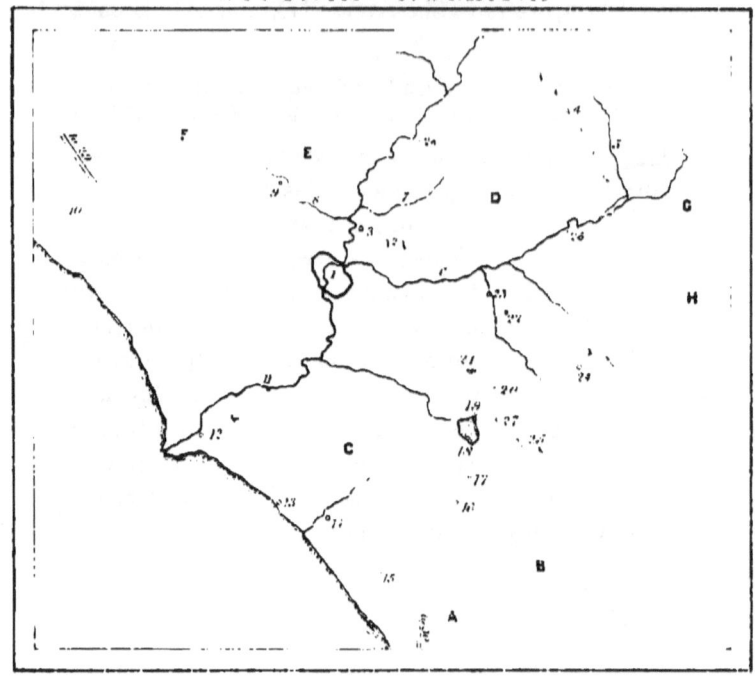

Roma (1).
Mons Sacer (2).
Fidenæ (3).
Lucretilis (4.)
Digentia (5).
Anio (6).
Allia (7).
Cremera (8).
Veii (9.)
Cære (10).
Tiberis (11).
Ostia (12).
Laurentum (13).

Lavinium (14).
Ardea (15).
Corioli (16).
Aricia (17).
Lacus Albanus (18).
Alba Longa (19).
Tusculum (20).
Lacus Regillus (21).
Gabii (22).
Collatia (23).
Præneste (24).
Algidus (25).
Tibur (26).

Mons Albanus (27).
Crustumerium (28).
Tarquinii (29).
Antium (30).
Rutuli (A).
Volsci (B).
Latini (C).
Sabini (D).
Veientes (E).
Etrusci (F).
Equi (G).
Hernici (H).

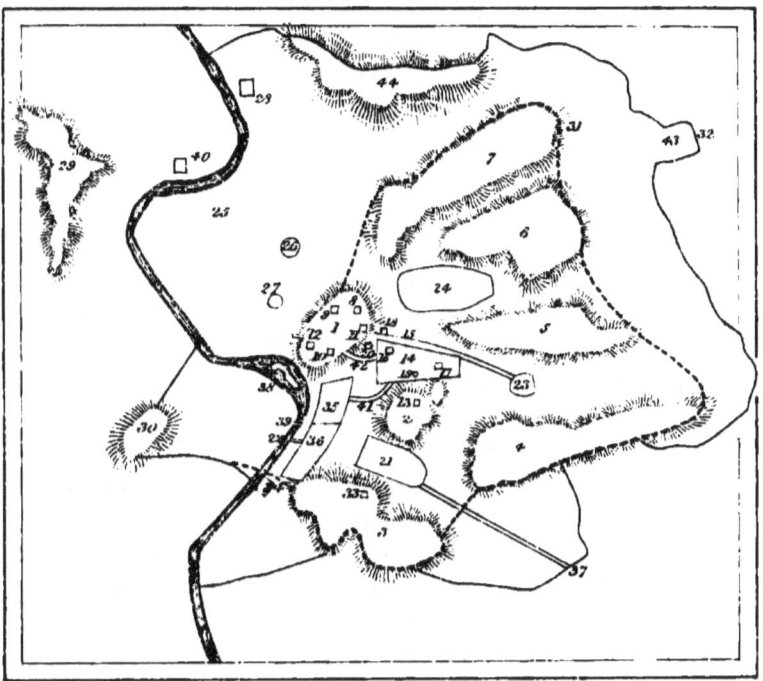

ROMA

Capitolinus (1).
Palatinus (2).
Aventinus (3).
Cœlius (4).
Esquilinus (5).
Viminalis (6).
Quirinalis (7).
Jupiter Capitolinus (8).
Jupiter Feretrius (9).
Jupiter Tonans (10).
Rupes Tarpeia (11).
Arx (12).
Palatium Augusti (13).
Forum (14).
Via Sacra (15).

Rostra (16).
Milliarium Aureum (17).
Curia (18).
Tullianum (19).
Templum Concordiae (20).
Circus Maximus (21).
Cloaca Maxima (22).
Colosseum (23).
Suburra (24).
Campus Martius (25).
Pantheon (26).
Theatrum Pompeii (27).
Mausoleum Augusti (28).
Vaticanus (29).

Janiculum (30).
Agger Servii Tullii (31).
Agger Aureliani (32).
Juno Regina (33).
Emporium (34).
Velabrum (35).
Forum Boarium (36).
Porta Capena (37).
Insula Tiberina (38).
Pons Sublicius (39).
Moles Hadriana (40).
Vicus Tuscus (41).
Clivus Capitolinus (42).
Castrum Prætorium (43).
Pincius (44).

C

ROME was built on seven hills, viz., **Palatinus**, the original abode of Romulus, who added the **Capitolinus, Cœlius,** and **Quirinalis.** The **Aventinus,** added by Ancus Martius, and the **Viminalis,** and **Esquilinus** by Servius Tullius. On the N.E. of the Capitolinus stood the temple of Jupiter Capitolinus; of Jupiter Feretrius, where the *spolia opima* were deposited. On the S.W. was the **Arx** and the temple of Jupiter Tonans, etc. On the E. was the **Tarpeia Rupes,** whence criminals were flung. The **Palatinus** contained the **Casa Romuli** and **Ficus Ruminalis;** but it was almost entirely occupied in later times by the Palace of Augustus and the temple of **Apollo,** with its splendid library. Between the two hills was the **Forum,** the great centre of Roman life, bounded: on the N. by the **Via Sacra,** along which processions went to the capitol: on the E. by the **Comitium,** where public business was transacted. In the Forum stood the **Rostra:** the columna **Rostrata;** the **milliarium aureum,** whence distance was measured. Adjacent to the **Forum** were numerous temples, *e.g.,* **Vesta, Castor, Pollux,** the **Curia,** or Senate House; on the N., the Tullianum, or lower dungeon; on the N.W., the temple of **Concord,** in which the Senate sometimes met. Between the Palatinus and Aventinus was the **Circus Maximus** ($\frac{1}{2}$ mile long), the great race-course of Rome, capable of containing 385,000 people. Westward was the Forum Boarium, or market, and the Velabrum. It was here that the **Cloaca Maxima,** or great sewer, discharged its contents into the Tiber, still in perfect order after the lapse of so many years. Lower down was the **Emporium,** or docks, the busiest part of the city. Between the Cælian and Esquiline was the **Colosseum,** began by Vespasian, capable of holding

87,000 spectators. The **Suburra** (24), the slums of Rome, lay in the valley between the Quirinal, Viminal, and and Esquiline. The **Campus Martius**, originally an open plain for gymnastics, warlike exercises, and meetings of the people, became in course of time covered with numerous buildings, among which we may mention the **Pantheon** of Agrippa; **Theatrum Pompeii**, in the Curia or large hall, of which Cæsar was assassinated; the **Mausoleum** of Augustus. The **Vicus Tuscus** (41), leading from the **Forum** to the Velabrum, was the fashionable shopping street in Rome.

GRAECIA.*

Graecia (the Roman name) and **Hellas**-*ădos* (the Greek) are terms which, taken in their widest acceptation, comprehend **Peloponnesus, Graecia Propria, Thessalia, Epirus,** and **Macedonia.** If we add to the last-named the contiguous country of **Thracia,** the whole will present a portion of the earth's surface (extending between the parallels of Lat. 36° and 41° N. and the lines of Long. 19° and 27° E.) which may be regarded as forming an irregular triangle, with the mountain chain of **Haemus** (1), for its base, the coast lines of the Aegean and Ionian Seas for its sides, and for its apex Cape **Taenarus** (2), (Matăpan), the southern extremity of the *Peloponnesus,* and of Greece. This triangular space is nearly bisected by the **Pindus** (3) Chain, which forms the water-shed of the whole territory of Greece, separating the rivers on its Eastern side which flow into the Aegean, from those on the Western side which flow into the Ionian Sea.

I.
Peloponnesus, (Morea).

The localities of prime classical interest in the leaf-shaped peninsula called **Peloponnesus** which forms the southernmost division of Greece are the following:

Among the Mountains, which cover a large portion of the surface, are, 1. **Cyllene** (4), fabled to have been the

* Haec cuncta Graecia, quae famâ, quae gloriâ, quae doctrinâ, quae plurimis artibus, quae etiam imperio et bellicâ virtute floruit, parvum quendam locum in Europâ tenet, semperque tenuit.— *Cicero pro Flacco,* 27.

GRÆCIA ET INSULÆ 1-147

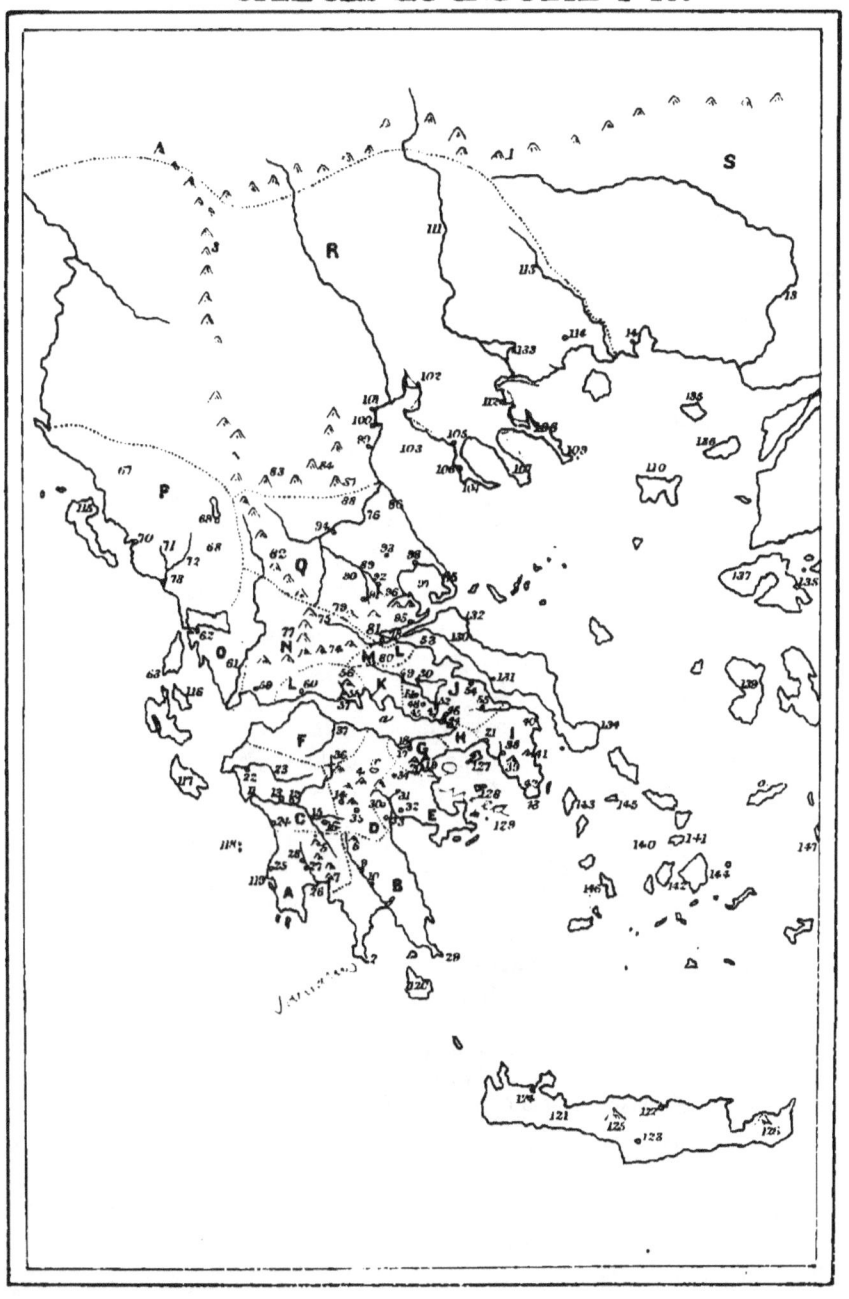

spot where Mercury was born and his stepping-stone between heaven and earth, when, acting as messenger of the gods ('Deorum nuncius') he either lighted upon, 'the heaven-kissing hill,' or re-ascended from it:—2. and 3. **Lycaeus** (5), and **Maenalus** (6), the favourite haunts of *Pan* (ovium custos) :—4. **Taygetus** (7), the resort of the Spartan maidens ('virginibus bacchata Lacaenis')—a range of mountains, now called Pentedactylon, which bounds on the W. the basin of **Eurotas** (8). On that river stood **Lacedaemon** *v* **Sparta** (9), so long the rival of Athens, not in arts, but in arms. And lower down, **Amyclae** (10), the home of Tyndarus, Castor and Pollux, and the scene of the Hyacinthian festival. The only other river in the peninsula worth noting here, is **Alpheus*** (11), on which, not far from its embouchure, was the town of **Pisa** (12); and near it the plain of **Olympia** (13), where the Olympic games were celebrated by all Greece in the first month of every 5th year, each intervening period being called an Olympiad. Among the mountains where *Alphēus* rises stood **Mantinea** (14), the field of the second great victory gained by the Theban Epaminondas over the Spartans, and the scene of his death, B.C. 362. On the **Helisson** (15), a tributary of *Alphēus*, stood **Megalopolis** (16), b.-pl. of the historian **Polybius**, and of **Philopoemen**, 'the last of the Greeks.'

The other localities in the Peleponnesus worth recording here will be best learned in connection with the *six* little Departments (*five* maritime, and *one* inland), into which it was divided: viz. **Achaia** (F), bounded on the N. by

---

* ——— that renownèd flood, so often sung,
  Divine Alphèus, who by secret sluice
  Stole under seas to meet his Arethuse.—*Milton, Arcad.* 29.

Sinus Corinthiacus (a), (Gulf of Lepanto), Elis (c), Messenia (A), Laconia (B), Argolis (E), and, inland, Arcadia (D).

1. In Achaia, on the *Isthmus*, was Corinthus (17), (poeticè, *Ephyre*), destroyed by Mummius, B.C. 146, and rebuilt by Caesar. It had a port on each side of the Isthmus, Lechaeum (18), on the Corinthian Gulf, and Cenchreae (19), on the Saronic: and hence the epithet *bimaris*. The citadel was on the summit of a rock called *Acrocorinthus*, where sprung the fountain Pirene (20). To the east of the Isthmus stood Eleusis (21), celebrated for the Eleusinian Mysteries in honour of Ceres, and as being the b.-pl. of Æschylus, the tragedian.

2. In Elis, besides Pisa, Olympia, and Elis (22), on the Peneus (23), was Pylos (24), thought by some to have been the city of Nestor, the veteran sage of the Iliad.

3. In *Messenia* was another Pylos (25), (Navarino); and in the basin of the stream Pamisus (26), Messene (27), whose citadel Ithome (28), was called by Philip of Macedon 'one of the horns of the Peloponnesus,' Acrocorinthus being the other.

4. On the *Laconian* coast were two promontories, Taenarus (2), noted as one of the passages to the infernal regions, 'Taenarias etiam fauces, alta ostia Ditis,' and Malea (29) v. *Malēa*, a cape dangerous to mariners.

5. In *Argŏlis* were Argos (30), a favourite city of Juno, called by Homer ἱππόβοτον, which Horace translates *aptum equis*:—Mycenae, (31), the city of Agamemnon:—Tiryns (32, the reputed b.-pl. of Hercules, who is thence called 'Tirynthius heros:'—Lerna (33), and its Marsh, the abode of the many-headed Hydra, which it was one of the twelve labours of Hercules to destroy; and Nemea

(34), the haunt of the Nemean lion, the killing of which was another of these labours.

6. In the only *inland* department **Arcadia**, besides *Mantinēa* and *Megalopōlis* already mentioned, were also the very ancient city of **Tegea** (35), hard by Tripolítzà, the modern capital of the Morea; **Erymanthus** (36), the mountain haunted by the boar which Hercules slew: and **Stymphalus** (x), a lake infested by birds (*Stymphalīdes*) which fed on human flesh, till they were destroyed by Hercules. Still further north, on the confines of Achaia, was the celebrated **Styx** (37), one of the rivers of the lower world.

II.
Graecia Propria.

The Isthmus of Corinth, embracing **Corinthia** (G), and **Megaris** (H), connects *Peloponnesus* with **Graecia Propria**, the notable localities of which will be best indicated by referring each to the Ancient Division in which it was situated. Of these divisions, the principal were, **Attica** (I), **Boeotia** (J), **Doris** (M), **Locris** (L), **Phocis** (K), **Aetolia** (N), and **Acarnania** (O).

1. In **Attica** stood **Athenae** (38), 'the eye of Greece, Mother of Arts and Eloquence,' distinguished by her triple harbour. *Piraeus* (Πειραιεύς), *Munychia* and *Phalērum*, by her *Acropōlis* and its *Parthĕnon*, and by her 'School of Ancient Sages' viz., 'the olive-grove of *Academe*, Plato's retirement:'—'*his* too, who bred Great Alexander to subdue the world;'* and 'painted *Stoa* next.'† There too were '*Ilisssus*' whispering stream,'

* *Lyceum*, the School of Aristotle and the Peripatetics.
† The School of Zeno and the Stoics.

and 'Flow'ry hill **Hymettus** (39), with the sound Of bees' industrious murmur.'* To the East of Hymettus was **Poeania**, the b.-pl. of the orator Demosthenes, and **Colonus**, the b.-pl. of Sŏphocles, and to the East of Athens, was the plain of **Marathon** (40), so famous for the defeat of the Persians by Miltiades, B.C. 490; Mt. **Pentelicus** (41), (Mendeli) which furnished marble for the building of the Parthĕnon; the silver mines of **Laureon** (42); and the southern promontory **Sŭnium** (43), crowned with the temple of Minerva Sunias, the pillars of which still remaining give name to the modern Cape Colonne.

2. Of **Boeotia**, the "cockpit" of Greece, the low country was proverbial for its thick atmosphere and the *pingue ingenium* of its inhabitants; but the mountains, **Cithaeron** (44), and **Helicon** (45), with its fountain **Hippocrene** (45), and the heights which enclose the plain, were all of a character so widely different, that, under the common name of *Aonia*, they were celebrated by the poets as the haunts of the Muses, who were thence called *Aonĭdes* and *Aoniae puellae*. In Boeotia were the Towns of **Thebae** (46), the capital, b.-pl. of Epaminondas and Pindar; south of it **Platacă** (47), where the confederate Greeks, commanded by Pausanias, defeated the Persians under Mardonius, B.C. 479; where also the inhabitants were cruelly put to death by the Lacedaemonians, B.C. 427; and **Leuctra** (48), where Epaminondas gained his first victory over the Lacedaemonians, B.C. 371. On the north, near the mouth of the **Cephissus** (49), was **Chaeronea** (50), memorable (1.) for the defeat of Athenians by Baeotians, B.C. 447; and, (2.) for the

* See the whole passage quoted from, in Par. Reg. iv. 238.

defeat of Baeotians by Philip, which put an end to the liberties of Greece, B.C. 338. South-east of this was **Coronea** (51), the scene of Agesilaus' victory over the Thebans and their allies, B.C. 394. Below it, **Ascra** (52), b.-pl. of Hesiod (Ascreo seni). On the narrow strait called **Euripus** (53), which separates Boeotia from the island Euboea, was **Aulis** (54), where the Grecian fleet destined for Troy was detained by contrary winds, till Agamemnon consented to the required sacrifice of his daughter Iphigenia; and further down, **Tanagra** (55), the scene of many conflicts between the Athenians and Spartans, B.C. 457.

3. Of **Phocis** the remarkable features are, 1. The fountain-head and early course of the **Cephissus** (49), whose lower basin forms the northern portion of Boeotia. 2. Mt. **Parnassus** (56), with its double top, ($διχόρυφος$, bicornis), 'mons Phoebo Bromioque sacer.' Between the two peaks was **fons Castalius** (56), and farther down, on the **Pleistus** (57), of which the Castalian spring is a feeder, stood the temple of Apollo, in which were the Tripod of the Pythia and the seat of the famous Oracle of **Delphi** (58).

4. **Aetolia** was known in early Greek story as the country ravaged by the Calydonian Boar, till it was slain at last by Meleager. The boar got its name from **Calydon** (59), the city of Tydeus and his son Diomede, the latter so well known to the reader of Homer and Virgil under his patronymic title $Τυδείδης$, Tydides. East of Calydon, and on the Sinus Corinthiacus, was **Naupactus** (60), ($ναῦς$-$πήγνυμι$), a celebrated naval station. **Achelous** (64), the longest and largest of Grecian rivers, forms the boundary between Aetolia and

5. **Acarnania,** a district which lies between that river and the Ambracian Gulf. It was at the entrance of this gulf, near the promontory **Actium** (62), that the naval battle was fought between Augustus and Mark Antony, which secured to the former the undisputed sovereignty of the Roman Empire, (B.C. 31). Lower down, on an island, was **Leucate*** (63), the 'lover's leap' of the ancient world, whence Sappho threw herself into the sea.

III.
Epirus.

Between the Ambracian Gulf and the Acroceraunian Promontory, lay the extensive region of **Epirus** including **Chaonia** (67), and **Molossia** (68). It was famed for its breed of horses, and of Molossian dogs, and still more so for the most ancient of all the Greek oracles **Dodona** (69). On the coast was **Sybota** (70), where the naval battle was fought, B.C. 432, which occasioned the Peloponnesian War. A little to the south are the celebrated streams **Cocytus** (71), (κωκυω) and **Acheron** (72), (αχος-ῤεω), which fall into the **Acherusia Palus** (73), Milt. P. L. ii. 573. √

Having now reached the *western* limits of Greece we return eastward to the coast of the Aegean, at the point where it trends Northward from the boundary of *Graecia Propria,* and crossing, or winding round, **Mt. Oeta** (74), which is an offset Eastward from the *Pindus* chain, we find ourselves in the country called by the ancients √

*   Pete protinus altam
Leucada, nec saxo desiluisse time.—*Ovid Heroid*, xv. 165.

Thessalia.

Physically considered, **Thessaly** is made up of the Basins of two rivers, the **Spercheos** v. **Sperchius** (75), and the **Peneus** (76), (Σπερχειὸς and Πηνειὸς). The *Sperchius* rises in **Mt. Tymphrestus** (77), one of the heights of that Pindus range whence so many streams 'dispart to different seas.' It flows Eastward into the **Sinus Maliacus** (78), (Gulf of Zeitoun) through a *broad valley* (Scoticé *strath*, Graecé αὐλών, Gallicé *bassin*,) which is bounded by two ranges of hills, offsets from Pindus, **Mt. Othrys** (79), on the N. and on the S. **Mt. Oeta** (74).* At the Eastern extremity of **Mt. Oeta** is the famous pass called **Thermopylae** (80), where Leonidas and his 300 Spartans made their bold stand against Xerxes, B.C. 480; and near it **Anticyra** (81), noted for producing hellebore, which the ancients looked upon as an antidote against madness. 'Tribus Anticyris caput insanabile' is said by Horace of a person incurably deranged; or, if the case

---

* The vale of the *Sperchius* must have had great natural beauty, to have been selected by Virgil, in the following exquisite lines, as one of the retreats which a lover of rural scenery would delight to dwell in:—

> Rura mihi et rigui placeant in vallibus amnes,
> Flumina amem silvasque inglorius! O, ubi campi,
> Sperchēosque, et Virginibus bacchata Lacaenis
> Taÿgeta! * O qui me gelidis in vallibus Haemi
> Sistat, et ingenti ramorum protegat umbra!
>   *Georg.* ii. v. 485.

* It is not unusual to call *Cyllene* the highest hill in the Peloponnesus, and *Taenarus* the most southern point of Europe. But the latest researches have made it out, that one of the Five Fingers, (πίντι δακτύλων) as the *Taÿgeta* are now called, is 116 feet higher than *Cyllene*; and that Tarifa, the Southern extremity of Spain, is 23 m. farther South than *Taenarus*, and is consequently the most Southern point of Europe.

is not hopeless, 'Naviget Anticyram.' There is an Anticyra in the Corinthian Gulf, which is often confounded with this.

The other and by much the larger portion of Thessaly is the basin of the **Peneus**. It is a territory containing 4000 square miles of surface, and possessing the singular property of being encompassed on all the *four* sides, even the side facing the sea, by ranges of mountains; on the W. by **Pindus** (82); on the N. by **Montes Cambunii** (83), and **Pierii** (84); on the S. by **Othrys**; and on the E. by the range of **Pelion** (85), **Ossa** (86), and **Olympus**, (87), the three hills, by the piling of which, one upon the other, the giants attempted to scale the heavens. To the continuity of this mountain barrier—the lips as it were of the great basin—there is but one interruption; and it consists of a rent in the rocky barrier between Olympus and Ossa, through which **Peneus**, the single main river of Thessaly Proper, finds its way to the Aegean. The outlet bore the name of **Tempe** (88). It is a valley which in some places is so narrow as barely to allow the river to pass between the opposite cliffs. This fact, coupled with the general aspect of the country, which presents to the eye an interminable plain, has led to the almost unavoidable conclusion, that Thessaly Proper was once a vast Lake, and furnished no land for the habitation of man till the rent at Tempé was either formed, or so deepened as to admit the efflux of the waters produced in the summits and inner slopes of the enclosing heights. It was then, and not till then, that these waters found an issue in the one stream of the **Peneus**, which receives, incorporates, and discharges them all.

Among the numerous tributaries of the *Penēus*, one

that joins it on the *right* called **Apidanus** (89), is worth noting, for two reasons: 1*st*, Because near the source of its tributary, the **Enipeus** (90), stood **Thaumaci** (91), (Θαύμακοι, 'the city of wonderment,' from θαυμάζω, miror,) so called because the traveller who has been toiling across Othrys first beholds here, *with astonishment*, the rich and to his eye boundless plain that stretches before him; 'repente,' says Livy, 'velut vasti maris expanditur planities, ut subjectos campos terminare oculis haud facile queas,' (B. xxxii. c. 4); And 2*dly*, Because, half way down the **Apidanus**, where it is joined by its feeder **Enipeus**, lies the field where the battle of **Pharsalia** (92), was fought between Caesar and Pompey, B.C. 48. North-east of Pharsalia was **Cynoscephalae** (93), Dog's Head, where Flaminius gained his celebrated victory over Philip of Macedon, B.C. 197.

On the **Peneus** itself, below the point where the **Apidanus** falls into it, stood **Larissa** (94), which some describe as the City of Achilles: but that honour belongs to another **Larissa** (95), not within the limits of the great basin, but in the S.E. corner of Thessaly called **Phthiotis**, the country of the **Dolopes** and **Myrmidones**. To this **Larissa** is often added, for distinction's sake, the epithet **cremaste**, *i.e.* **pensilis**, because it *hangs*, as it were, on the slope of that stripe of land which lies between the Eastern margin of the great basin and the Aegean. To the north of **Larissa** is the **Amphrysus** (96), where Apollo kept the herds of Admetus—hence called 'Pastor ab Amphryso.' On the **Pegasœus Sinus** (97), was **Iolcus** (98), the residence of Jason, the leader of the Argonautic expedition.

Proceeding, in our tour of the Mediterranean and its

cognate waters, from the North-eastern boundary of Thessaly, we find ourselves, for a great part of the journey, in countries where the classical interest is mainly confined to the line of coast, their interior having been either imperfectly known to the ancients, or seldom alluded to in their extant writings. Hence it is, that in these countries there is little worthy of note, in so brief an outline as this, beyond the sea-board on which the Greeks planted colonies and where the Greek language was spoken. Even

V.
Macedonia,

which we admitted as a fifth among the great Divisions of Greece, may be treated according to this rule.

Beginning, then, in our progress northward, at the 40th parallel of Lat. we find, along the Macedonian coast, **Pydna** (99), where Perseus was baffled in his last effort to save his kingdom from Roman dominion, B.C. 168. A little north was **Methone** (100), where Philip gained his first victory, B.C. 360. Farther north, on a Lake 15 miles from the sea, was **Pella** (101), the capital of Macedon, and b.-pl. both of the 'Vir Macĕdo,' Philip, and of 'Philip's warlike son' the 'Pellaeus Juvenis,' Alexander the Great, B.C. 356. Pursuing again the line of coast, we come to **Thessalonica** (102), at the head of **Sinus Thermaicus** (103), (Gulf of Salonichi). To the Christians of this city St Paul addressed his two 'Epistles to the Thessalonians.' We fall in next with three peninsular projections, 1. **Pallene** (104), on the isthmus of which stood **Potidaea** (105), and a little to the N. **Olynthus** (106), places whose names are familiar to the readers of

Demosthenes; 2. **Sithonia** (107),* and 3. **Acte** (108), across the isthmus of which, 12 stadia broad, Xerxes, we are told, cut a canal for a passage to his fleet. At the south end of *Acte* is **Athos** (109), a mountain so lofty, that, according to Pliny, it projects its shadow on **Lemnos** (110), (87 miles distant) when the sun is setting at the summer solstice. As we approach the mouth of the **Strymon** (111), at one time the boundary of Macedonia, we find **Stagira** (112), b.-pl. of Aristotle, who is hence called the 'Stagirite.' On the Strymon itself is **Amphipolis** (133), at the siege of which (B.C. 422) Brasidas, the brave Spartan, and Cleon, the Athenian leather-seller, were slain. In the country that lies between the rivers *Strymon* and **Nestus** (113), at some distance from the sea, was the field of **Philippi** (114), where the decisive battle was fought between Octavius (afterwards Augustus) and Mark Antony on the one side, and Brutus and Cassius on the other, B.C. 42.

ISLANDS OF GREECE.

Of the 'Isles of Greece' which ought to be familiar to every reader of the Classics, some are in the Ionian Sea, off the Western side of the Greek continent; but the great majority are on its Eastern side and in the Aegean.

I. On the *western* side are 1. **Corcyra** (115), (Corfu), thought to be the Homeric *Scheria* the island of the Phaeacians, in which the poet places the Gardens of Alcinöus: 2. **Ithaca** (116), the home of Ulysses, which,

---

* *Sithonius* is used by Horace and Ovid as synonymous with Thracian: thus 'Memphin carentem Sithoniâ nive.'—*Hor.*; and 'Brachia Sithoniâ candidiora nive.'—*Ov.*

though '*in asperrimis saxulis, tanquam nidulum affixam*,' he preferred to immortality in the brighter island of Calypso, (Cic. de Or. i. 44) : 3. **Zacynthus** (117), (Zante) *nemorosa*, as Virgil calls it, a colony from which is said to have peopled and given name to Saguntum : 4. Off the W. coast of Peloponnesus, the rocks called **Strophades** (118), the haunts of the Harpies—(*Aen.* iii. 210.) : 5. Opposite, but close to Pylos in Messenia, was the celebrated island of **Sphacteria** (119), where the noblest of the Lacedaemonians were captured, B.C. 425, (Thucyd. Bk. iv. 8, κ.τ.λ). To the S. of the Laconian promontory *Malea* was **Cythera** (120), an island sacred to Venus ; still farther south is **Creta** (121), of old ἑκατόμπολις ; but of its 'hundred cities' the only three known to fame in classical times were **Gnossus** (122), the seat of Minos, **Gortyna** (123), and **Cydonia** (124), all three famed for archery. Of its mountains, **Ida** (125), was the loftiest, and on **Dicte** (126), Jupiter is said to have been reared, and fed upon honey and the milk of the goat Amalthēa.

II. Of the Islands lying to the *east* of Greece and in the Aegean—after noticing **Salamis** (127), b.-pl. of Ajax, Teucer, Solon, Euripides, and memorable for defeat of Persians by Themistocles, B.C. 480, and below it **Ægina** (128), the 'eyesore' of the Piræus, both in the **Sinus Saronicus** (129),—let us visit first those worthy of mention which are situated to the North of the 38th parallel of Latitude. These are,

1. **Euboea** (130), an island stretching 93 miles along the coast of Boeotia and Attica, and approaching so near the continent in the channel called **Euripus** (53), as to admit of a bridge being thrown across. On this channel was the chief city of the island, **Chalcis** (131), (Negropont),

nearly opposite to *Aulis* in Bœotia. On the north is the promontory of **Artemisium** (132), whence the Greek fleet was obliged to retire before the Persians just before the battle of Thermopylae, B.C. 480. In doubling **Caphareus** (134), a promontory at the S.E. extremity of Euboea, the Grecian fleet on its return from Troy was overtaken by a storm, which partly destroyed and partly dispersed it. What the Greeks suffered on their way home, (says Diomede, one of the sufferers,)

———— scit triste Minervae
Sidus, et *Euboicae* cautes, ultorque *Caphareus*.
—*Aen.* xi. 260.

and Ovid, in allusion to the same disaster, says,

Quicunque Argolicâ de classe Capharea fugit
Semper ab Euboicis vela retorquet aquis.—*Trist.* i. 1. 83.

2. **Samothrace** (135), where the Corybantes, the priests of Cybĕle, practised the rites and mysteries of that goddess.

3. **Lemnos** (110), the island on which Vulcan alighted when

thrown by angry Jove
Sheer o'er the crystal battlements of heaven.

So sings Milton, as Homer had sung before him,

'Ρῖψι ποδὸς τιταγὼν ἀπὸ βηλοῦ θεσπεσίοιο.*

4. **Tenedos** (136), an island in sight of Troy, and, small as it is, not to be omitted, since Virgil pronounces it 'notissima famâ Insula.'

5. **Lesbos** (137), 'Insula nobilis et amœna,' with its capital **Mitylene** (138), near the Asiatic coast, the b.-pl. of Arīon, and of the lyric poets Alcaeus and Sappho.

6. **Chios** (139), (Scio), one of the seven places which

* Par. Lost, i. 740, and Iliad, i. line 591.

contended for the honour of giving birth to Homer. Its claim is admitted by Lord Byron when he calls him
> The blind old man of Scio's rocky isle.

7. **Samos** (a), famous for its commerce, b.-pl. of the philosophers Pythagoras (B.C. 540), and Epicurus (B.C. 342).

III. The numerous islets in the Aegean whose latitudes are lower than 38° ('crebris fretra consita terris'), are generally classed under two denominations:

1. The **Cyclades** (140), a group that clusters ($\grave{\epsilon}\nu\ \varkappa \acute{\upsilon}\varkappa\lambda\varphi$) round **Delos** (141),—that floating island which Neptune fixed with his trident, to be a resting-place for the persecuted Latona to bring forth Apollo and his twin sister Diana. Here was *Mt. Cynthus:* hence Cynthius Apollo. Among the most noted of this group are 1. **Paros** (142), famed for its statuary marble, and b.-pl. of Phidias, the sculptor, who made the noblest use of it: 2. **Ceos** (143), off the promontory of **Sunium** (43), b.-pl. of the elegiac poet Simonides; **Naxos** (144), an island that figures in the history of Bacchus and Ariadne; **Gyaros** (145), v.-*rae* and **Seriphos** (146), places of banishment for Roman criminals under the Empire.

2. The islets to the East of the Cyclades, from the circumstance of their being *scattered*, were called **Sporades** (147), from $\sigma\pi\epsilon\acute{\iota}\rho\omega$, *spargo*. They extended as far as, and included, *Icaria*, which took its name, as did the sea round it, from the fabled fate of Icarus, the son of Daedalus,—
>——— mersus in alto
> Icarus Icariis nomina fecit aquis.—*Ov.*

Returning from the Islands of the Aegean to the Eastern limit of Macedonia, we cross the *Nestus*, and find ourselves in THRACIA.

ATHENÆ 1--28

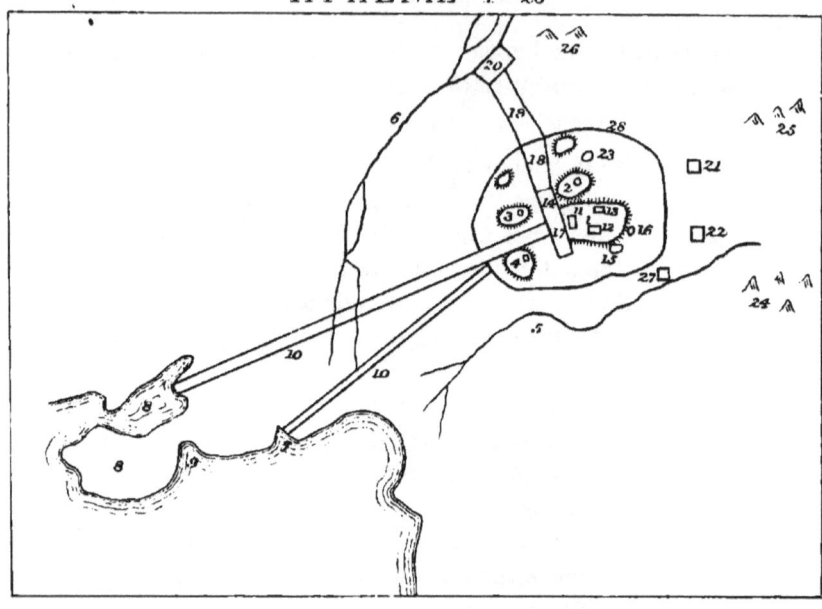

Acropolis (1).
Areopagus (2).
Pnyx (3).
Museum (4).
Ilissus (5).
Cephissus (6).
Phalerum (7).
Piraeus (8).
Munychia (9).
Longwalls (10).

Propylaea (11).
Parthenon (12).
Erectheum (13).
Agòra (14).
Theatrum Bacchi (15).
Odenm (16).
Prytaneium (17).
Ceramicus (inner) (18).
Ceramicus (outer) (19).
Academia (20).

Cynosarges (21).
Lyceum (22).
Templum Thescos (23).
Hymiettus (24).
Lycabettus (25).
Colonus (26).
Olympicum (27).
Walls of Themistocles (28).

ATHENS, like Rome, was built on what are called hills ($4\frac{1}{2}$ miles from the sea), the chief of which was the **Acropolis** (1), (150 feet high), with flat summit (1000 feet by 500). To the west of this was the **Pnyx** (3). To the N.W. the **Areopagus** (2). To the S.W. **Museum** (4). The Acropolis was central. On the S.E. and N. it was inaccessible. On the west it was entered by the **Propylaea** (11). The chief buildings within the enclosure were the **Parthenon** (12), or Temple of Athena, built of Pentelic marble in the purest Doric style: the **Erectheum**, the most revered building in Athens, and connected with its most ancient legends. The next hill is **Areopagus**, in the open air of which, and on benches hewn out, the **Upper Council** held its meetings. **Pnyx** (3), where the assembly of the people was held. **Theatrum Bacchi**, S.E. of the Acropolis was large enough to hold all the people of Athens. The seats were hewn out of the solid rock. Here the Greek Drama, etc., was represented. **Odeum** (16), built by Pericles, with a roof to imitate Xerxes' tent. **Prytaneium**, or **Tholus**, where the Prytaneis took their meals, and where foreign ambassadors were entertained. The Prytaneis were a sort of Committee of the Council of 500. The **Ceramicus** (inner) had the **Agora** at the S., in the depression between the Acropolis, Areopagus, and Museum. It had several stoæ or colonnades, in one of which (στοα ποικιλη) Zeno the Stoic taught. The **Ceramicus** (outer), where those who had a public funeral were buried. The **Museum**, or Temple of the Muses. Outside Athens was the **Academia**, the school of Plato. **Cynosarges**, the school of Antisthenes, and the **Cynics**. The **Lyceum**, the school of Aristotle and the **Peripatetics**. The **Templum Theseos** (23), N. of Areopagus, built of Pentelic marble, contained the bones of Theseus, brought by Cimon from Scyros. The **Olympieum** (27), the temple of the great Zeus (354 feet by 171), began by Pisistratus, was 700 years in building. It was of white marble, with columns $6\frac{1}{2}$ feet in diameter, and 60 feet high.

THRACIA,*

A country, the coast of which extends from the **Nestus** (113), along the shore of the *Aegean*, the *Hellespont*, the *Propontis*, the *Thracian Bosporus*, and the *Euxine* Sea, as far as **Mt. Haemus** (1), which was its northern boundary. Tracing the sea-board from the *Nestus* Eastward we come first to **Abdera** (14), which, though proverbial for the stupidity of its inhabitants, was nevertheless the b.-pl. of Democritus—a sage who showed more wisdom in laughing at the follies, than his brother philosopher of Ephesus did, in weeping over the vices, of mankind, (*Juv.* x. 28).

Farther East, we reach the mouth of **Hebrus** (13), a river on whose banks the poets feigned that Orpheus was torn in pieces by Bacchants and his head thrown into the stream. Few rivers out of Italy and Greece are more frequently alluded to in the classics than the *Hebrus* (now Maritza).

Next comes the Thracian peninsula called **Chersonesus** (19), on whose eastern side is **Hellespontus** (21), a strait separating Europe and Asia, and so narrow at one part as to have been swum across from **Abydos** (55), to **Sestos** (20),—by Leander in the fabulous ages of Greece, and in the present century by Lord Byron. Here Xerxes built his bridge of boats. A few miles east from Sestos was **Ægos Potamos** (22), where the Athenian fleet was utterly

* For the first two sections see Map of Greece (Thracia).

THRACIA,1 14: ASIA MINOR,1 61: SYRIA,1 16: ÆGYPTUS,1 9.

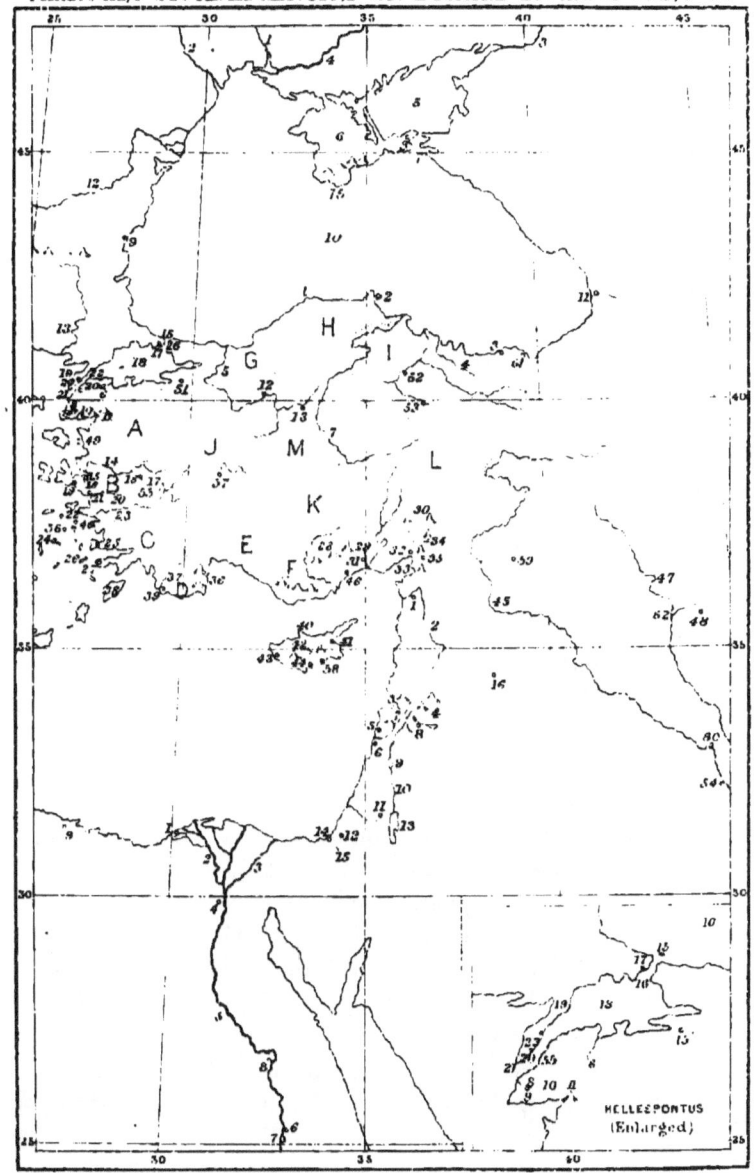

defeated by Lysander, B.C. 405, which put an end to the Peloponnesian War. The Hellespont widens into the sea-lake called **Propontis** (18), and, at the side of **Byzantium** (17), contracts again into the **Bosporus Thracicus** (16), (Strait of Constantinople), which, after keeping the two continents narrowly asunder for a considerable space, opens out again into **Pontus Euxinus** (10), (the Euxine or Black Sea). At the northern extremity of this Bosporus where it widens, are situated some rocky islets, which seemed to Jason and his Argonauts, as they approached them through the windings of the strait, to move and come into collision with each other, and were hence named **Symplegades** (15), (from συμπλήσσω, concutio). These rocks were also called *Cyaneae* (κυάνεαι), from their dark purple colour in the offing.

It is here that we commence our tour of the Black Sea. So, turning to the *left*, that we may continue, as we have all along done, to keep our *right* shoulder to the sea, we come first upon the site of **Tomi** (9), the place of Ovid's banishment, where he wrote his Fasti and Tristia, and where he ended his days. As we go North we encounter the mouths of **Danubius** (12), or *Ister*, on the *north* side of whose vast basin lived the *Daci*, a brave people, formidable to the Empire even under Augustus, ('conjurato descendens Dacus ab Istro'). On the *south* side, between *Haemus* and the Danube, was the country called *Moesia* (Servia and Bulgaria). Along the dreary shore N. of the mouths of the Danube, where lived the barbarous tribes *Getae, Gelōni*, and *Sauromătae* v. *Sarmatae* (the Scythians belonged rather to the lofty tableland of Central Asia), we meet with no object of greater classical interest than the embouchures successively of the

Tyras (2), the **Hypanis** (1), and the **Borysthenes** (4), till we come to **Chersonesus Taurica** (6), now called Crimēa, a peninsula somewhat larger than the Peloponnesus, and, on the south side, of considerable beauty and fertility. Next comes **Palus Maeotis** (5), 'the Tauric Pool,' now Sea of Azoff, formed by the influx of **Tanais** (3), the Don, and connected with the Euxine by the **Bosporus Cimmerius** (7), (Strait of Yenicale), on which was **Panticapaeum** (8), capital of Mithridates' Bosporic Empire, (now Kertsch).

Proceeding Eastward from this strait, we come in sight of the 'Hyrcanian cliffs of Caucāsus, and dark Iberian dales.'* These are in the country that lies between the Euxine and Caspian Seas, which many ethnologists regard as the primitive seat of the finest type of the human species. Travelling along the shore of the Euxine we reach, at its farthest point E., the river and town of **Phasis** (11), the city of *Aeetes* (Αἰήτης) king of *Colchis* and father of *Medēa*, names intimately connected with the myth of the Argonautic expedition under Jason, the date of which is reckoned anterior even to the Trojan war. The importation into Europe of the pheasant, *Phasiana* (sc. avis), is thought to have been one of the fruits of this expedition.

* Milton, Par. Reg. iii. 317.

Asia Minor.

We now travel back S.W. and then W. from the *Phasis* to the Thracian Bosporus whence we started. The points of interest on the Southern shore of the Euxine (*i.e.* on the N. coast of the peninsula of ASIA MINOR), are the following : 1. **Trapezus** (61) *-untis*, which, under the name of Trebizonde, was a city of great note, under the Eastern Empire ; 2. **Cerasus** (3), a town whence Lucullus transplanted into Italy the tree which retains its name in Latin, and appears in various corrupted forms in modern tongues, 'kersch,' 'cérise,' 'cherry ;' 3. the mouth of the **Thermodon** (4), the basin of which river was assigned as their dwelling-place to the fabled race of female warriors called Amazons (a priv. and $\mu\alpha\zeta\delta\varsigma$, mamma) ; 4. the river **Halys** (7), eastern boundary of the Lydian kingdom of Croesus, the crossing of which proved fatal to him in his contest with Cyrus king of Persia, ('et Croeso fatalis Halys,' *Lucan.* iii. 272) ; 5. **Sinope** (2), b.-pl. of Diogĕnes the Cynic ; 6. the promontory **Carambis** (1), opposite to **Kriumetopon** (15) (Cape Aia, near Balaclava) in the Crimea. The distance across is 150 miles; and this is the narrowest breadth of the Euxine; and 7. the mouth of the **Sangarius** (5), in the basin of which river, near its source in *Galatia*, were the towns of **Ancyra** (13), (Angora) famed for its goats' hair, and **Gordium** (12), for the oracular knot, see *Curtius*, B. iii. cap. 2 and 3.

The above localities, excepting the last two, are in the provinces called **Pontus** (I), **Paphlagonia** (H), and **Bithynia** (G). The other *maritime* provinces of that peninsula were in number six, *three* on the Asiatic shore of the Aegean, viz., 1. **Mysia** (A), including *Phrygia Minor* and the **Troad**; 2. **Lydia** (B) v. **Maeonia**, including **Ionia**, which was the sea-board of **Lydia** and thickly planted with Greek colonies; and 3. **Caria** (C), including the district of *Doris;* and *three* on the Mediterranean; 4. **Lycia** (D); 5. **Pamphylia** (E), including *Pisidia;* and 6. **Cilicia** (F). In all these six there are localities with whose names and positions every student ought to be made familiar. For example:—

1. In **Mysia**, it is sufficient to name, 1. **Troja** v. **Ilion** (8), built on an eminence between the **Simois** (10), and **Scamander** (9), overlooked by **Mt. Ida** (11), and itself overlooking the Plain of Troy; and 2. the river **Granicus** (6), on whose banks Alexander the Great gained his first victory over the Persians, (B.C. 334). Off *Mysia* were the **Arginusae Insulae** (49), where the Lacedaemonians were completely defeated by the Athenians, B.C. 406, who ungratefully put their generals to death, (*Xen. Mem.* B. 1, cap. 1).

2. In **Lydia** flowed the river **Hermus** (14), and its tributary **Pactolus** (17), both famed for the gold found in their sands; and on the Pactolus, at the foot of **Mt. Tmolus** (55), was **Sardis** (18), the capital of **Lydia**. S. of the *Hermus* was **Smyrna** (15), on the river **Meles** (16), one of the cities which contended for the honour of being the b.-pl. of Homer; and hence the poet has been called Melesigenes.*—On the Ionian sea-board were 1. **Teos** (19), b.-pl. of Anacreon; 2. The mouth of the river **Cayster**

---

* *Blind* Melesigenes, thence *Homer* called,
Whose poem Phoebus challenged for his own.
Milt. Par. Reg., iv. 529.

(20), famed for its swans,* where stood **Ephesus** (21), with its magnificent temple of Diana, set fire to and burnt to the ground the same day on which Alexander the Great was born; 3. **Mt. Mycale** (22), off which the Greeks gained a naval victory over the Persians, the very day (479 B.C.) on which Mardonius was defeated at *Plataea*. We next cross the **Maeander** (23), a river of great length, and so remarkable for its windings that it has furnished an English verb descriptive of a similar character in all other streams.† South of the Maeander, but still to be reckoned an Ionian city, was **Miletus** (24): most of the Greek colonies that fringed the borders of the Euxine Sea were planted by the Milesians.‡ Miletus was noted also for its wool ('Milesia magno Vellera mutantur,' *Virg. G.* iii. 306), and as the b.-pl. of Thales, the earliest and not the least sagacious of the Greek philosophers. South west of Miletus, but at some distance from the coast, was **Patmos** (56), the scene of St John's banishment.

3. On the coast of **Caria** stood **Halicarnassus** (25), a city rendered memorable (ἀξιόλογος) as the b.-pl. of the two great historians Herodotus and Dionysius, and also for the sepulchral monument of Mausōlus, reared by his queen Artemisia, the name of which has passed into a

---

* Ad vada Caÿstri concinit albus olor.—*Ov.*

† The word is still more widely and metaphorically used as an appellative in Latin. Thus, Cicero, in Pison. c. 22, says,—'Quos *macandros* quaesisti?' What quirks, subterfuges, or evasions, have you had recourse to? See also Virg. Aen. v. 251.

‡ Ovid, writing from *Tomi*, says,—

Huc quoque Mileto missi venere coloni,
Inque Getis Graias constituere domos.
Ov. Trist. iii. 9, 3.

household word in our own and other modern tongues.* On the opposite side of the bay stood **Cnidos** (27), where was a statue of *Venus* ('Regina Cnidi Paphique,' *Hor. Od.* i. 30.) reckoned the master-work of Praxitĕles; and at the entrance of this bay, midway between *Halicarnassus* and *Cnidus* lay the island **Cos** (26), b.-pl. of the famous physician and medical writer Hippocrătes, and of Apelles the most celebrated of Grecian painters.† Cos was noted also for its wines, and for the manufacture and dyeing of fine cloth.‡ Off the coast of Caria is another island much larger and more noted than Cos, viz., **Rhodos** (38), Rhodes, in the capital of which, of the same name, was the brazen statue of the Sun called *Colossus*, 70 cubits high, which bestrode the entrance of the harbour.

4. Moving eastward, along the Carian Shore, we enter **Lycia**, and having crossed the **Xanthus** (37), arrive at **Patara** (39), the winter residence according to the poets of Apollo, as **Delos** (141), (*materna*) was his favourite in summer.§

5. In **Pamphylia**, the point of greatest interest is **Climax** (36), a spur of the *Taurus* range, which comes so abruptly and perpendicularly upon the shore that Alexander's army marched under the cliff breast high in

* Mausolēum.

† Apelles is said never to have spent a day without employing his pencil: hence the proverb, 'Nulla dies sine lineâ.'—See Plin. Nat. Hist. xxxv. 12.

‡ Horace (*Od.* iv. 13, 13) says to a faded beauty:—

    Nec Coae referunt jam tibi purpurae,
    Nec clari lapides tempora, quae semel,
        Notis condita fastis, Inclusit volucris dies.

§ Delius ac Patărcus Apollo.—*Hor. Od.* iii. 46. 4.

the water of the Mediterranean, which had been driven shore-ward by an easterly wind.

6. **Cilicia** extends from the eastern limit of Pamphylia to the **Sinus Issicus** (33), and Mt. **Amanus** (30), and has the mountain chain of **Taurus** (28), for its northern boundary. The western portion of Cilicia is *rough* and hilly hence called *Trachēa* (τραχεῖα, *aspera*); the eastern, more level and fertile, is called *Campestris* (πεδιάς). On the coast of the latter, as we approach the river **Cydnus** (29), we pass through **Soli** (46), (Σόλοι), a place worthy of mention here, only because like Μαίανδρος and Μαυσώλειον, it has furnished the English language with a word.* We then come to **Cydnus** (29), the river that so nearly proved fatal to Alexander the Great; and ascending it we arrive at **Tarsus** (31), the capital of the province, and classed by Strabo—himself a native of Asia Minor,—with Athens and Alexandria, as a seat of art, science, and refinement. *Tarsus* was the b.-pl. of St Paul. The last town in Cilicia was **Issus** (32), situated at the head of the bay named from it, **Sinus Issicus** (33). It was here that Alexander gained (B.C. 333) his second great victory over the Persians, and made prisoners of war the wife, mother, and infant son of Darius, (B.C. 333). In this neighbourhood were the *Pylae* **Amanicae** v. *-ides* (34), and **Pylae Syriae** (35), narrow passes or gorges in the mountain range *Amānus* which runs N.E. from the bay of Issus till it joins Mt. *Taurus*.

* At *Soloi* a colony from Athens settled, whose Attic Greek degenerated so notoriously, that any Athenian who violated at home the purity and propriety of Attic speech was said σολοικίζειν to *solœcise*, and his offence was called Σολοικισμός. Of the two words we have scarcely adopted more than the noun, and we apply it to other things besides language, as, when we say, 'a *solœcism* in politics.'

CYPRUS AND ITS TOWNS.

All these localities are frequently mentioned, both in the history of Alexander's Expedition into Asia, and in Cicero's account of his proconsulate in Cicilia as given in his own Letters and Despatches.

7. In **Pontus** we meet with **Amasea** (52), b.-pl. of Mithridates and Strabo the Geographer. Above it **Zela** (53), where Caesar defeated Pharnaces, son of Mithridates (B.C. 47), announcing the victory in his well known despatch, 'veni, vidi, vici.'

8. In **Bithynia** we have **Nicæa** (51), where the Nicene Creed was drawn up under Constantine the Great, A.D. 325.

The *inland* provinces of Asia Minor were **Phrygia Magna** (J) (in which is **Synnada** (57), noted for its marble), **Galatia** (M), **Cappadocia** (L), and **Lycaonia** (K).

Fronting the *Sinus Issicus* is **Cyprus** (40), the favourite island of Venus, and hence epithets are applied to the goddess derived from towns and temples there :—such as, *Cypria*, **Paphia** (43), **Idalia** (42), **Amathuntia** v. *Amathusia* (44), *Salaminia*. The last epithet is taken from **Salamis** (41), a town at the east end of Cyprus, said to have been founded by Teucer, who, not being allowed by his father Telamon to land on his native **Salamis** (127), in the Saronic gulf, upon his return from the Trojan war, was assured by the oracle of Apollo

Ambiguam tellure novâ *Salamina* futuram.—*Hor. Od.* i. 7, 29.

On the S. was **Citium** (58), the b.-pl. of Zeno, and the place where Cimon the Athenian died, B.C. 449.

The range of Mt. *Amānus* forms the S.E. boundary of ASIA MINOR, separating it from **Syria**, in like manner as, on the N.E., the river **Euphrates** (45), separates it from *Armenia*.

TOWNS IN THE EAST.

The Asiatic peninsula thus bounded—of which we have merely traced the sea-board of low rich land skirting the shores of the Euxine, Propontis, Aegean, and Mediterranean Seas—is a country little inferior in size to the Spanish peninsula, to which, in many respects, it bears a resemblance. Both lie between the same parallels of Lat. (36° to 43°), and of both a vast extent of table-land occupies the interior.

We may here mention a few places of great note, still more to the east: (1.) **Carrhae** (59), on a tributary of the Euphrates, where Crassus was defeated, B.C. 53, by the Parthians: (2.) **Arbela** (48), where the decisive battle was fought between Alexander and Darius, B.C. 331, which ended the Persian Kingdom. Near this place, but on the Tigus, was **Ninus** (62), or *Nineveh*, a very ancient city, destroyed, B.C. 625, by the Medes and Persians: (3.) **Babylon** (54), one of the most celebrated cities in the ancient world. Here Alexander died, B.C. 323. A little to the north was **Cunaxa** (60), where Cyrus was defeated and slain by Artaxerxes, B.C. 401: (4.) at the head of the Persian Gulf was **Persepolis**, destroyed by Alexander the Great in a fit of drunken frenzy.

The TRACT of land which bounds the Mediterranean to the East, from the 31st to the 37th degree of N. Latitude, was called Syria.

Syria.

It comprehended **Phoenicia, Palaestina,** and **Judaea,** and extended Southward from Mt. Amānus to the confines of Egypt. The physical peculiarity of this region is a chain of Mountains running parallel with the shore of the Mediterranean, and never very far distant from it. The highest part of the chain is the Lebanon of Scripture, at the point where it diverges into two branches called **Libanus** (3), **Antilibanus** (4). These two enclose the Basin of the river **Leontes** (7), a portion of the country, which, from its physical aspect, the Greeks call Κοίλη Συρια (i.e. *cava*). *Coelesyria*, then, is an appellation corresponding in name and nature to the Scotch word '*how*' (i.e. hollow), as when we speak of the 'How o' the Mearns.'

The only other Syrian rivers worth noticing are,—

1. The **Orontes** (2), which, rising in the same elevated ground as the *Leontes*, flows in the opposite direction *northward*, till at last it makes a bend westward, and about 20 m. from the sea passes the famous city **Antiocheia** (1), the capital of Syria; in the vicinity of which, was

> that sweet grove
> Of Daphne, by Orontes.—*Par. Lost*, IV. 270.

2. **Jordanes** (10), the Jordan, which rises on the South side of Lebanon, and flowing nearly due South forms in its course the **Waters of Merom**, the **Lake Tiberias** (9), in Scripture the Lake of Gennesareth or Sea of Galilee. The river disappears at last in the **Lacus Asphaltites** (13),

the Dead Sea,—a bituminous lake without issue.* Halfway between this *Mare Mortuum* and the Mediterranean, on the brook Kedron, stood **Hierosolyma**, (11), JERUSALEM, the metropolis of Palestine.

Having dwelt hitherto on the interior aspects and localities of Syria, we now resume our journey along the Coast, proceeding southward from the mouth of the *Orontes*. Before reaching the mouth of the *Leontes*, we come upon **Sidon** (5), and, after crossing that river, upon **Tyros** (6). Both of these cities are in **Phoenicia**, and were the earliest, most enterprising, and wealthiest of all ancient states. On the same parallel of Latitude as Sidon, but eastward beyond the chain of *Antilibanus*, was the

delightful seat
Of fair **Damascus** (8), on the fertile banks
Of Abāna and Pharphar, lucid streams.—*P. L.* i. 467.

Still farther E., in what is now a desert of sand, was the splendid city of **Palmyra** (16), (Tadmor), historically associated with the name of Queen Zenobia and her secretary Longinus, author of the Treatise on the Sublime.

The Southern Division of Syria was **Palaestina**, to which and to its sub-divisions *Galilaea*, *Samaria*, and *Judaea*, are attached recollections and associations of an interest higher and more sacred than the classical, and which it would be foreign to my purpose to touch upon here.

The last town of any note along the coast as we approach Egypt, was **Gaza** (12), which had a port on the Mediter-

* Nec Jordanes pelago accipitur; sed unum † atque alterum lacum integer perfluit; tertio retinetur.—*Tacit. Hist.* v. c. 6.

† This first lake is *named* by no ancient author but Josephus, who calls it *Samachonitis*, probably the 'Waters of Merom' of Scripture.

ranean, and was a flourishing place till it was sacked by
Alexander the Great. Milton, though he speaks of
'Gaza's frontier bound,' was not unaware that Palestine
extended farther S. even to the brook called **Torrens
Aegypti** (15), and **Rhinocolura** (14), at its mouth. Here
the coast begins to trend Westward, and we travel over
a portion of the dreary desert of *Arabia*. Taking care to
shape our course so as to steer clear of 'that Sirbonian
bog, Where armies whole have sunk,' (*P. L.* i. 592), we
reach at last the Eastern branch of **Nilus**, the Nile, the
river, and the only river of Aegyptus.

AEGYPTUS,

A country forming the N.E. portion of the great peninsula or continent of AFRICA, lying between the 23° 30' (Tropic of Cancer,) and 32° 15' N. Lat., and 30° to 35° E. Long.

There is perhaps no part of the world, out of Italy and Greece, to which the poets and orators of antiquity make more frequent allusion than to Egypt; yet no ancient writer who is not a professed geographer goes much into detail, or mentions more than one or two of its towns and localities. The singular nature of the country, the immemorial existence of the Pyramids, the dim tradition of of a very remote antiquity, the absence of rain, the periodical inundations and mighty cataracts of the River, and, above all, the now explored, but as the ancients thought, unexplorable fountain-head of the **Nile** (5), which the river god studiously concealed from mortals,—all these circumstances threw a charm of sublimity and interest over the whole, which captivated the imagination both of the poet and his readers : And hence the frequent question, so strikingly put by Tibullus (i. 7. 23.) when he asks,—

NILE PATER, quânam possum Te dicere causâ,
Aut quibus in terris occuluisse caput ?

But the sculptural and architectural remains of unknown date, which modern research has brought to light at Luxor, Carnac, the island of **Philae** (7), and elsewhere, do not

seem to have been fully appreciated by, or even known to, the ancients. Of the Towns and Villages so thickly planted on the banks of the Nile, (Herodotus, iii. 177, states the number at 10,040 !) none have a claim to be enumerated here, with the exception of the following:
1. **Syene** (6), (Assouan), a town so nearly under the tropic as to justify Lucan in saying (ii. 587), 'umbras nusquam flectente Syene,' meaning by the expression that at the summer solstice, when the sun is on the meridian, the shadow is not projected northward, as it is in all higher latitudes: 2. **Thebae** (8), (*Niloticae*), which must be regarded as one of the largest and most ancient of cities, seeing it is described by Homer as having a hundred gates (ἑκατόμπυλοι) and capable of sending forth from each of them 200 men-at-arms with chariots and horses: of all which, and of the city itself, not a vestige remained even in Juvenal's time,

Atque vetus Thebe centum jacet obruta portis.—*Sat.* xv. 6 ;

and, 3. **Memphis** (4), on the left bank of the Nile, with the Pyramids in its immediate neighbourhood. *Memphitica Tellus* is used as a poetical synonym for *Aegyptus*. Farther down, the Nile separated into different channels, by all of which its waters found their way to the sea. Of these channels the ancients enumerated seven: and hence the constant allusion among the poets to 'septemflua, septena, septemgemini, septemplicis *ostia Nili*.' The most noted of the seven were the outermost on either side, *Ostium* **Canopicum** (2), W. and **Pelusiacum** (3), E. These two diverging branches, with the sea-coast line between their mouths, form the sides and base of the triangular space Delta, so called from its resemblance to the capital form of that letter in the Greek alphabet; and it is by these two channels alone that the water of the Nile is now discharged. Twelve miles west from the Canopic embouchure was **Alexandria** (1), (so named after Alexander the Great, who founded it on his way back from the Oasis and temple of Jupiter Ammon), a great city in ancient times, as it is even now.

LIBYA—NORTHERN COAST OF AFRICA.

As we advance westward from *Alexandrīa* we pass **Paraetonium** (9), the frontier town of Egypt, S. of which was the Oasis of Jupiter Ammon. After this we meet with nothing to detain us on the coast of **Libya** till we reach **Cyrene** (61),* the b.-pl. of Aristippus, Carneades, and Callimachus. In the latter days of Greece *Cyrēne* was a flourishing colony, where Art and Philosophy were cultivated; but at the present day, *etiam periêre ruinae*. Nor is there anything to redeem the desolation of this region as we proceed, till we pass successively the **Syrtus, major** (59) **and minor** (60). The latter terminates in **Lacus Tritonis** (61), a locality connected obscurely with the mythological history of Minerva, who is often called *Tritonia Virgo*. From this point commences a region of great natural fertility, which was long the 'granary' of Rome, and is rich in historical recollections.

First, we have **Africa Propria**, the proper domain of **Carthago** (56), the great rival of **Roma**; and 27 m. along the coast, **Utica** (55), where the second Cato rather than submit to Caesar, put a period to his life; and hence he is distinguished from Cato Major by the epithet *Uticensis*.

* See Map of Italy.

In the interior is **Zama**, where the elder Scipio Africanus defeated Hannibal. On the east coast is **Thapsus** (58), memorable for the victory gained by Caesar, B.C. 47, over Metellus Scipio and Pompey's party.

We then enter *Numidia*, the country of Jugurtha, and the scene of the first exploits of Marius, which prepared the way for Metellus *Numidicus* to finish the war and carry Jugurtha prisoner to Rome.

The last western division of this African coast was *Mauretania*, the kingdom of Bocchus and of Juba, bounded on the N. by the Mediterranean, on the W. by the Atlantic, and on the S. by the lofty range of Mt. **Atlas**, which protects it from the encroachments of the ocean of sand that lies beyond. As we approach the Atlantic, we come in sight of **Abyla** (4),* (Rock of Ceuta) and **Calpe** (Rock of Gibraltar), the two Pillars of Hercules, on the opposite sides of the *Fretum Herculeum:* And thus we have completed our tour of the Mediterranean, and have reached at last

—— longae FINEM chartaeque viaeque.—*Hor. Sat.* i. v. 104.

* See Map of Spain.

INDEX.

	Page		Page
Abdera,	54	Amphrysus,	46
Abydos	54	Amyclae,	38
Abyla,	71	Anas,	3
Acarnania,	40, 43	Andes,	27
Achaia,	38	Ancona,	26
Achelōus,	42	Ancȳra.	58
Acheron,	43	Anio,	20
Acherusia Palus,	43	Anticȳra,	44
Acroceraunia,	31	Antilibanus,	65
Acrocorinthus,	39	Antiocheia,	65
Acte,	48	Antium,	23
Actium,	43	Aonia,	41
Addua,	27	Apidănus,	46
Adula,	13	Apulia,	22
Aedui,	14	Aquae Sextiae,	14
Aegates Insulae,	30	Aquileia,	30
Aeoliae Insulae,	28	Aquīnum,	20
Ægina,	49	Aquitani,	14
Ægos Potamos,	54	Arar,	13
Ægyptus,	68	Arbela,	64
Aetna,	29	Arcadia,	39
Aetolia,	40, 42	Arduenna,	15
Agrigentum,	29	Arelate,	13
Agylla v. Caere,	23	Arethusa,	29
Alesia,	12	Arginusae,	40
Alexandria,	70	Argolis,	39
Allia,	20	Argos,	39
Allifae	20	Ariminum	26
Allobroges,	15	Arnus,	19
Alpes,	13	Arpinum,	20
Alpes Maritimae,	16	Arsia,	28, 31
Alphēus,	38	Artemisium,	50
Amanus Mt.,	62	Arverni,	8
Amasea	63	Ascra,	41
Amathus,	63	Asculum,	21
Amiternum,	21	Aternus,	19
Amphipolis	48	Athenae,	40

	Page		Page
Athesis,	28	Canusium,	21
Athos,	48	Caphareus,	50
Atlas,	71	Cappadocia,	63
Attica,	40	Capua,	20
Aturus	10	Carambis,	58
Aufidus,	19	Caria,	59
Augusta Taurinorum,	26	Carrhae,	64
Aulis,	42	Carthago,	70
Aureliani,	11, 14	Casilinum,	21
Avenio,	13	Castulo,	4
		Catino,	29
BABYLON,	64	Caucasus,	57
Baetis,	3	Caÿster,	59
Baiae,	24	Celtae,	14
Baleares Insulae	7	Cenchreae,	39
Barcino,	7	Ceos,	51
Basilia,	12	Cephissus	41
Belgae,	14	Cerasus,	58
Beneventum,	21	Chaeronea,	41
Bilbilis,	5	Chalcis,	49
Bithynia,	58	Chaonia,	43
Boeotia,	40	Charybdis,	29
Borysthenes,	57	Chersonesus Taurica,	57
Bosphorus Cimmerius,	57	Cher. Thracia,	54
Bos. Thracicus,	56	Chios,	50
Brundusium,	25	Cilicia,	59
Bruttii,	22	Cithaeron,	41
Burdegala,	11	Citium,	63
Byzantium,	56	Clanis,	20
		Climax,	61
CAECUBUS,	23	Clusium,	20
Caere,	23	Cnidos,	61
Caesar Augusta.	5	Cocytus,	43
Cajeta,	23	Coelesyria,	65
Calabri,	25	Colchis,	57
Calagurris,	5	Colonia Agrippina,	12
Calenus,	23	Colonus	41
Cales,	21	Confluentes,	12
Calle,	3	Corinthus,	39
Calor,	21	Corcyra,	48
Calpe,	6, 71	Corduba,	4
Calydon,	42	Corfinium,	21
Cambunii Montes,	45	Coronea,	42
Campania,	22	Corcica,	28
Campus Spartarius,	4	Corunna,	5
Cannae,	21	Cos,	61
Canopicum Ostium,	69	Cremera.	20

INDEX.

	Page		Page
Cremona,	26	Eurotas,	35
Creta,	49	Euxinus Pontus,	56
Croton,	25		
Cumae,	23	FAESULAE,	19
Cunaxa,	64	Falernus Ager,	23
Cyaneae,	56	Fescenninum,	20
Cyclades,	57	Fibrenus,	20
Cydonia,	49	Fidenae,	20
Cydnus,	62	Florentia,	19
Cyllēne,	36	Fons Bandusiae,	21
Cynocephalae,	46	Fons Castalius,	42
Cyprus,	63	Formianum,	23
Cyrene,	70	Fretum Herculeum,	6
Cythera,	49	Fretem Siculum,	24
		Furcae Caudinae,	21
DACI,	56		
Damascus,	66	GADIR v. GADES.	6
Dalmatia,	31	Galatia,	63
Danubius v. Ister,	56	Galēsus,	24
Daunia,	22	Gallia Comāta,	14
Delos,	51, 61	Gallia Transpadana,	22, 27
Delphi,	42	Gallia Cispadana,	22, 27
Dicte,	49	Galilaea,	66
Digentia,	20	Gargānus,	25
Dodona,	43	Garumna,	8, 10
Dordogne,	11	Gaza,	66
Doris,	40	Gebenna,	8
Drepanum,	30	Geloni,	56
Drilo,	31	Genabum,	11
Druentia,	13	Genua,	22
Dubis,	14	Getae,	56
Durius,	3	Gnossus,	49
Dyrrhacium,	25	Gordium,	58
		Gortīna,	49
ELEUSIS,	39	Graecia,	36
Elis,	39	Granicus,	59
Emerita Augusta,	4	Gyaros,	51
Enipeus,	46		
Enna,	30	HAEMUS,	36
Ephesus,	60	Halicarnassus,	60
Epirus,	36	Halys,	58
Erymānthus,	40	Hebrus,	54
Eryx,	30	Helicon,	41
Etruria,	22	Helisson,	38
Euboea,	49	Hellespontus,	54
Euphrates,	54	Helvetii,	8
Euripus,	42, 49	Heraclēa,	24

INDEX.

	Page		Page
Herculaneum,	24	Lacus Trasÿmenus,	19
Hermus,	59	Lacus Tritonis,	70
Hicrosolỹma,	66	Lacus Verbanus,	27
Himĕra,	30	Larissa,	46
Hippocrene,	41	Larissa Cremaste,	46
Hispalis,	4	Latium,	22
Hydruntum,	25	Laurentum,	23
Hymettus,	41	Laurĕon,	41
Hypanis,	57	Lavinium,	23
		Lechaeum,	39
IBERUS,	3	Lemnos,	48, 50
Icaria,	51	Leontes,	65
Ida Mt.,	49, 59	Lerna,	39
Idalia,	54	Lesbos,	50
Ilerda,	5	Leucate,	43
Ilissus,	40	Leuctra,	41
Ilion v. Troja,	50	Libănus,	65
Illyricum,	31	Liburnia,	31
Ilva,	28	Libya,	60
Interamna,	26	Liger,	10, 11
Iolcus,	46	Liguria,	22
Isara,	15	Lilybaeum,	29
Issus,	62	Lingones,	12, 14
Istria,	22	Liris,	19
Italia,	15	Locris,	40
Italica,	4	Lucania,	22
Ithaca,	48	Lugdunum,	13
Ithome,	39	Luna,	22
Itius Portus,	15	Lunensis Portus,	22
		Lutetia Parisiorum,	12
JAPYGIUM,	24	Lycaeus,	38
Jordanes,	65	Lycaonia,	63
Judaea,	65	Lycia,	59
Jura,	8	Lydia,	59
KRIUMETŌPON,	58	MACEDONIA,	36
		Macra,	17
LACEDAEMON,	38	Maeander,	60
Lacinium,	25	Maenalus,	38
Laconia,	39	Maeotis Palus,	57
Lacus Albanus,	23	Malea,	39
Lacus Asphaltites,	65	Mautinea,	38
Lacus Benācus,	27	Mantua,	27
Lacus Brigantinus,	12	Marathon,	41
Lacus Larius,	27	Mare Inferum,	19
Lacus Lemanus,	13	Mare Superum,	19
Lacus Regillus,	23	Marsi,	22

INDEX.

	Page		Page
Massicus, Mt.,	23	Nola,	24
Massilia,	14	Nova Carthāgo,	6
Matrŏna,	12	Numantia,	3
Mauretania,	71	Numidia,	71
Medoacus Minor,	28		
Megalopolis,	38	OCTODURUS,	13
Megaris,	40	Oeta Mt.,	43
Meles,	59	Olympia,	38
Melita,	30	Olympus,	45
Memphis,	69	Olynthus,	47
Messenia,	39	Olysipo,	3
Messēne,	39	Orontes,	65
Metapontum,	24	Ortygia,	29
Metaurus,	26	Ossa,	45
Metellinum,	4	Ostia,	20
Methone,	47	Othrys,	44
Miletus,	60		
Mincius,	27	PACHȲNUS,	29
Minius,	3	Pactolus,	59
Minturnae,	20	Padus,	17, 26
Misenum,	27	Paestum,	24
Mitylene,	50	Pallēne,	47
Moesia,	47	Palmyra,	66
Molossia,	43	Palaestina,	65
Mons Sacer,	20	Paludes Pomptinae,	23
Morĭni,	15	Pamīsus,	39
Mosa,	10, 12	Pamphylia,	59
Mosae Trajectus,	12	Panormus,	30
Mosŭla,	12	Panticapaeum,	57
Munda,	6	Paphlagonia,	58
Munychia,	40	Paraetonium,	69
Mutina,	27	Parissii,	14
Mycăle,	60	Parnassus,	42
Mycēnae,	39	Paros,	51
Mylae,	30	Parthenŏpe,	24
Mysia,	59	Patăra,	61
		Patavium,	28
NAMNETES,	11, 14	Patmos,	60
Naupactus,	42	Peligni,	22
Naxos,	51	Penticlicus Mons,	41
Neapolis,	22	Pelion,	45
Nemĕa,	39	Pella,	47
Nervii,	15	Peloponnesus,	36
Nestus,	48	Pelorus,	29
Nicæa,	63	Pelusiacum Ostium,	69
Nilus,	67	Perusia,	19
Ninus,	64	Persepolis,	64

	Page		Page
Penēus,	39, 44	Saguntum,	6
Phalerum,	40	Salamis,	49, 63
Pharsalia,	46	Saldŭba,	5
Phasis,	57	Salo,	5
Philippi,	48	Samăra,	10
Philae,	68	Samaria,	57
Phocis,	40	Samos,	51
Phœnicia,	65	Samothrāce,	50
Phrygia,	54	Samnium,	22
Phthiotis,	46	Sangarius,	58
Picenum,	22	Sardis,	65
Pierii Montes,	45	Sardinia,	28
Pindus,	36, 45	Sauromatae,	56
Piraeus,	40	Scaldis,	10
Pirēne,	39	Scamander,	59
Pisa,	30	Scylla,	25, 29
Pistoria,	19	Seduni,	13, 14
Plataea,	41	Segesta,	30
Pleistus,	42	Sena Gallica,	26
Pœania,	41	Sena Julia,	26
Pompeii,	24	Sentinum,	26
Pontus,	58	Sequana,	10, 12
Portus Magonis,	7	Sequani,	8, 14
Potidaea,	47	Serīphos,	51
Propontis,	56	Sestos,	54
Provincia Rom,	14	Sicoris,	5
Pydna,	47	Sicilia v. Trinacria,	29
Pylos,	39	Sidon,	66
Pylae Amanicae et Syriae,	62	Simatheus,	29
Pyrenaeus Mons,	7	Simois,	59
		Sinōpe,	58
Raudii Campi,	26	Sinus Cantabricus,	10
Ravenna,	27	Sinus Corinthiacus,	39
Rhegium,	25	Sinus Issicus,	62
Rhenus,	10, 12	Sinus Ligusticus,	22
Rhinocolura,	67	Sinus Maliacus,	44
Rhodanus,	10, 13	Sinus Pegasæus,	46
Rhodos,	61	Sinus Saronicus,	49
Roma,	20	Sinus Tarentinus,	24
Rotomagus,	12	Sinus Thermaicus,	47
Rubicon,	17, 26	Sirbonian Bog,	67
Rudiae,	25	Sirmio,	27
		Sithonia,	48
Sabātus,	19	Smyrna,	59
Sabis,	15	Soli,	62
Sabini,	22	Sparta,	38
Sacrum Promontorium,	6	Spercheos,	44

INDEX.

	Page		Page
Sphacteria,	49	Tiberis,	19
Sporades,	51	Tiberias,	65
Stagira,	48	Tibur,	20
Strongyle,	29	Ticinus,	27
Strophades,	49	Timavus,	28
Strymon,	48	Tiryns,	39
Stymphalus,	40	Tmolus,	65
Styx,	40	Tolosa,	10
Sucro,	3, 4	Tomi,	56
Sulmo,	21	Torrens Ægypti,	67
Sunium,	41, 51	Trapezus,	58
Sybaris,	24	Trebia,	27
Sybota,	43	Treveri,	15
Syene,	69	Tridentum,	28
Symplegades,	56	Troja v. Ilium,	59
Synnada,	63	Turia,	3
Syracusae,	29	Tymphrestus,	44
Syria,	56	Tyros,	66
Syrtis Major,	70	Tyras,	57
Syrtis Minor,	70		
		Umbria,	22
Tader,	3	Utica,	70
Taenarus,	36, 39		
Tagus,	3	Valentia,	4
Tanais,	57	Varus,	16
Tanagra,	42	Veii,	20
Tarentum,	24	Venafrum,	20
Tarraco,	7	Veneti,	11
Tarsus,	62	Venusia,	21
Taurus,	62	Venetia,	22
Taygetus,	38	Verona,	28
Tegea,	40	Vesulus, Mons,	26
Telamon,	22	Vesuvius,	24
Tempe,	45	Vesontio,	15
Tenedos,	50	Via Appia,	25
Teos,	59	Vogesus,	8
Tergeste,	31	Vultur,	21
Thapsus,	71	Vulturnus,	19
Thaumaci,	46		
Thebae,	41	Xanthus,	61
Thebae Niloticae,	69		
Thermodon,	58	Zacynthus,	49
Thermopylae,	44	Zama,	71
Thessalia,	36, 44	Zancle,	29
Thessalonica,	47	Zela,	63
Thracia,	36		

EDINBURGH:
PRINTED BY H. AND J. PILLANS.
12 THISTLE STREET.

Educational Works

PUBLISHED BY

ADAM AND CHARLES BLACK.
EDINBURGH.

SCHOOL ATLASES.

Black's School Atlas of Modern & Ancient, Physical and Scripture Geography, and the Elements of Astronomy: a Series of 41 Maps by W. HUGHES, F.R.G.S., and J. BARTHOLOMEW, F.R.G.S. Royal 4to or 8vo, 10s. 6d. Maps all coloured.

LIST OF MAPS, 1 to 40.

Physical Geography.

1. Chief Physical Features of the World. 2. Ethnography. 3. Zoology 4. Botany. 5. Comparative View of Mountains and Rivers.

Astronomy.

6. Northern Celestial Hemisphere. 7. Southern Celestial Hemisphere. 8. The Solar System. 9. Theory of the Seasons.

Modern Geography.

10. World in Hemispheres. 11. Europe. 12. England and Wales. 13. Scotland. 14. Ireland. 15. France in Departments. 16. France in Provinces. 17. Holland and Belgium. 18. Prussia and German States. 19. Austrian Empire. 20. Switzerland. 21. Italy. 22. Spain and Portugal. 23. Sweden, Norway, and Denmark. 24. Russia. 25. Turkey in Europe and Greece. 26. Asia. 27. Turkey in Asia, &c. 28. Hindostan, &c. 29. Africa. 30. North America. 31. United States and Canada. 32. West Indies. 33. South America. 34. Australia, New Zealand, &c. 35. British Empire.

Ancient Geography.

36. World as known to the Ancients. 37. Italia, Northern Part. 38. Italia, Southern Part. 39. Greece and Islands of Ægean Sea.

Scripture Geography.

40. Palestine, with its Ancient Divisions and the Peninsula of Mount Sinai. 41. Countries embraced within the Travels of St. Paul.

SCHOOL ATLASES—*continued.*

Black's Modern Atlas. A Series of Twenty-seven
Maps Coloured. With Preface by W. HUGHES, and an Index of 15,000 Names. In one volume, 4to, cloth, price 5s.

Black's School Atlas for Beginners. A Series of
Twenty-seven Coloured Maps. In one vol. square 12mo, 2s. 6d.

Pillans' Classical Geography, with Maps.

First Steps in the Physical and Classical Geo-
graphy of the Ancient World. By JAMES PILLANS, late Professor of Humanity in the University of Edinburgh. Tenth edition, 12mo, 1s. 6d.

The Tales of a Grandfather. (History of Scot-
land.) By Sir WALTER SCOTT, Bart. Part I. A.D. 1033-1596.—Macbeth to Queen Mary. Part II. A.D. 1600-1700.—James VI. to Queen Anne. Part III. A.D. 1701-1760.—The Union to the Highland Rebellion. Each Part, Sixpence. Or in Cloth Limp, Eightpence. The same, complete in one volume 8vo, cloth, price 2s. 6d.

Scott's Poems—Author's Edition. With all the
Author's Introductions, and Notes by J. G. Lockhart (which are *wanting* in all other editions.)

Lay of the Last Minstrel. Author's Edition.
Marmion. Author's Edition.
Lady of the Lake. Author's Edition.
Rokeby. Author's Edition.
Lord of the Isles. Author's Edition.
Bridal of Triermain. Author's Edition.

Price One Shilling each in cloth. The same in 8vo, sewed, double columns, price 6d. each.

Elasticity and Heat. By Sir WILLIAM THOMSON,
Professor of Natural Philosophy in the University of Glasgow. Reprinted from the Encyclopædia Britannica, 4to, cloth, price 6s. (May also be had separately, price 4s. each).

Lectures on the History of Education in Prussia
and England, etc. By JAMES DONALDSON, LL.D., Rector of the High School of Edinburgh. Crown 8vo, price 3s. 6d.

CLASS-BOOKS.

Class-Book of English Prose, comprehending
Specimens of the most Distinguished Prose Writers from CHAUCER to the Present Time, with Biographical Notices and Explanatory Notes. By Rev. ROBERT DEMAUS, M.A. 12mo, 4s. 6d. Also to be had in Two Parts. Part I., containing the Prose Writers from CHAUCER to SOUTH; Part II., ADDISON to the Present Time. Price 2s. 6d. each.

"A very excellent Class-Book. . . . Its specimens of English Prose extend from Chaucer to Ruskin, and great care and judgment are evinced in their selection; not only are they for the most part excellent and characteristic specimens of the style of their Author, but, wherever possible, they have been chosen so as to give a lively idea of the character of the writers, as well as of their mode of treating their subjects."—*Westminster Review.*

"A volume which the mere general reader may peruse with pleasure, and which the student of English Composition may consult with advantage."—*Notes and Queries.*

Class-Book of English Poetry; comprising Extracts
from the most Distinguished Poets of the Country. By DANIEL SCRYMGEOUR. Fifth Edition, 12mo, 4s. 6d. Also, to be had in Two Parts. Part I., containing the Poets from CHAUCER to OTWAY. Part II., PRIOR to TENNYSON. Price 2s. 6d. each.

Introduction to the History of English Literature.
By Rev. ROBERT DEMAUS, M.A. 12mo, cloth, 2s.

"We have been much pleased with an 'Introduction to the History of English Literature,' by Robert Demaus, M.A., which, though a comparatively small Manual, is of sufficient extent to give a very good notion of our literature from the earliest times to the present. The leading writers and works are ably described, and even of those more concisely treated enough is said to indicate their true character. It is not often that so much condensed information is conveyed in so lively and agreeable a manner."—*Athenæum.*

"Mr. Demaus is already favourably known as the author of more than one excellent educational work. His 'Class-Book of English Prose,' which comprehended specimens of the most distinguished prose writers from Chaucer to the present time, has been universally accepted as a valuable contribution to scholastic literature."—*Illustrated London News.*

M. Masson's (of Harrow) French Class-Books.

Class-Book of French Literature, Comprehending
specimens of the most distinguished writers from the earliest period to the beginning of the present century, with Biographical Notices, Notes, Synoptical Tables, and a copious Index, by GUSTAVE MASSON, B.A., M.R.S.L., Assistant Master at Harrow School, and Member of the Académie des Sciences, Arts, et Belles-lettres de Bordeaux. Crown 8vo (517 pages). Price 5s. 6d. bound in cloth.

Introduction to the History of French Literature.
By GUSTAVE MASSON, B.A., M.R.S.L., Assistant Master at Harrow School, etc. etc. 12mo, cloth, 2s. 6d.

"Excellently adapted for its purpose as a handbook for the upper classes of Schools."—*Westminster Review.*

Kunz and Millard's French Grammar.
Grammar of the French Language, founded upon
the principles of the French Academy, by JULES A. L. KUNZ, Teacher of Modern Languages in the Edinburgh Institution, and FREDERICK MILLARD, B.A. Oxon., Principal of the Grammar School and Government Inspector of Schools, St. Kitts. Second edition, 12mo. Price 3s. 6d.

LATIN AND GREEK.

Latin Reader of Jacobs and Classen. Edited, with
Notes and a Vocabulary, by JAMES DONALDSON, M.A., LL.D., Rector of the High School of Edinburgh. 12mo, 3s. 6d.

*** The above may be had in Two Parts, at 1s. 9d. and 2s. each. FIRST COURSE: Exercises on the Inflexions. SECOND COURSE: Fables, Mythology, Anecdotes, Roman History, etc.

"An excellent Latin Reading Book."—*Athenæum.*

"A thoroughly sensible and serviceable edition of a valuable manual."—*Literary Gazette.*

A Course of Exercises in Latin Prose Syntax,
adapted to Ruddiman's Rules, with Copious Vocabularies. Crown 8vo, 3s. 6d. Or in two parts. Part I.—Agreement and Government. Part II.—The Syntax of the Subjunctive Mood. By W. S. KEMP, LL.D, late Head Classical Master in the High School of Glasgow. Crown 8vo, cloth, Part I. 1s. 6d., Part II. 2s.

Introductory Latin Exercises. By W. S. KEMP,
LL.D., late Head Classical Master in the High School of Glasgow, etc., Crown 8vo, cloth, price 1s.

Elementary Grammar of the Greek Language.
By Dr. L. SCHMITZ, late Rector of the High School of Edinburgh. Second Edition, 12mo, 3s. 6d.

Exercises in Attic Greek for the use of Schools
and Colleges, by A. R. CARSON, LL.D., F.R.S.E., late Rector of the High School of Edinburgh. 12mo, 4s.

BOTANY, GEOLOGY, &c.

BALFOUR.

Manual of Botany; being an Introduction to the Study of the Structure, Physiology, and Classification of Plants. By J. H. BALFOUR, M.D., late Professor of Botany in the University of Edinburgh. Crown 8vo, pp. 700, 12s. 6d.

Elements of Botany for the use of Schools. By same author. Illustrated with 427 Wood Engravings. Fcap. 8vo, pp. 321. Price 3s. 6d.

Introduction to the Study of Palæontological Botany. By same author. With 4 Lithographic Plates and upwards of 100 Woodcuts. Demy 8vo, price 7s. 6d.

JUKES.

JUKES' School Manual of Geology. New Edition, edited by ALFRED J. JUKES-BROWNE, of St. John's College, Cambridge. Price 4s. 6d.

WILSON.

Elements of Zoology, for Schools and Senior Classes, by Dr. ANDREW WILSON, Lecturer on Zoology, Edinburgh. With numerous Illustrations, 12mo. Price 5s.

ARITHMETIC, ALGEBRA, & MATHEMATICS.

PROFESSOR KELLAND.

Elements of Algebra for the use of Schools and Junior Classes in Colleges. By Rev. PHILIP KELLAND, M.A., late Professor of Mathematics in the University of Edinburgh. Crown 8vo, 4s.

Algebra; being a complete and easy Introduction to Analytical Science. By the same Author. Crown 8vo, price 7s. 6d.

HUGO REID.

A First Book of Mathematics; being an easy and practical Introduction to the Study. For Middle and Elementary Schools. By HUGO REID. Fcap. 8vo, price 2s.

DR. BRYCE.

Algebra. By JAMES BRYCE, M.A., late of the High School of Glasgow. Fourth Edition. Crown 8vo, price 3s. 6d.

The Arithmetic of Decimals, adapted to a Decimal Coinage. With numerous Illustrations of Improved Modes of Reckoning. By JAMES BRYCE, M.A., LL.D., F.G.S. Second edition, enlarged. 1s. 6d.

Key to Algebra. Crown 8vo, price 5s.

READING-BOOKS.

Demaus's Elementary Reading-Book, on a new plan, and with the view of forming the habits and cultivating the feelings of the pupils (with illustrations), by Rev. ROBERT DEMAUS, M.A. 18mo, pp. 269. Price 1s. cloth.

Buchan's Advanced Prose and Poetical Reader. A Collection of Select Specimens of English Literature, with Explanatory Notes and Questions on each Lesson; to which are appended Lists of Prefixes and Affixes, and an Etymological Vocabulary. By ALEXANDER WINTON BUCHAN, F.E.I.S., Teacher, West Regent Street Academy, Glasgow. 12mo, cloth, 3s.

By the same Author.

The Poetical Reader, a New Selection of Poetry for the School-Room, with Notes and Questions. 12mo, cloth, 1s. 6d.

OSWALD'S ETYMOLOGICAL WORKS.

The Etymological Primer: Part First, containing the Prefixes, Postfixes, and several hundred Latin and Greek Roots of the English Language, by Rev. JOHN OSWALD. Nineteenth Edition. Price One Penny.

The Etymological Primer: Part Second, or an Abridgment of the Etymological Manual. Ninth Edition, paper cover. Price 6d.

**The Etymological Manual: containing the Prefixes, Postfixes, and Latin, Greek, and other Roots of the English Language, adapted to the Improved System of Education. Eighteenth Edition, limp cloth. Price 1s.

A Dictionary of Synonymes and Paronymes of the English Language, limp cloth. Price 1s. 6d.

Published by A. and C. Black.

An Etymological Dictionary of the English
Language, adapted to the Modern System of Tuition. Tenth Edition.
18mo, cloth. Price 5s.

Just published, in fcap. 8vo, cloth, price 3s. 6d.

History of Charles XII., King of Sweden. From
the French of VOLTAIRE. With Vignette Portraits of Charles, Peter
the Great, and Voltaire; and Two Maps, one showing the extent of
Sweden and Poland at the time of the history, and the other Sweden
of the present day. Accompanied by a Sketch of Charles' Character
by Mrs. GRANT of Laggan, and Index.

> "A good service is done to literature and to students of history by the republication of Voltaire's *Charles XII. of Sweden*, in its English form. The outcry which has been raised against Voltaire by people who in most cases knew nothing about him, has unquestionably prevented his works from being known to the extent that it is desirable they should be known. This is especially true of his life of Charles XII. That work is an effort to strip history of the fables and superstitious nonsense which, before Voltaire's time, were the almost invariable accompaniments of historical books. Unfortunately they are not yet wholly absent from what is called history; indeed, there are historians who refuse to look at facts that tell against the fables upon which they have set their heart. Voltaire, however, had no tendency in that direction. His fault was not that of believing too much. Yet that he had not a perfectly analytical faculty, this book is witness. There is no need to recommend it to those who have made history their study; but those who have not read it will do well to get this edition. It is neatly got up in a small octavo volume."—*Scotsman*.

RELIGIOUS INSTRUCTION.

BISHOP BROMBY

The Church Student's Manual. By the Right
Rev. C. H. BROMBY, D.D., Bishop of Tasmania, late Principal of the
New College, Cheltenham. 12mo, cloth, red edges, price 3s.

Book of Common Prayer. Separately, 1s. 4d.

Class-Book of Scripture History. By Rev. R.
DEMAUS, M.A., etc. Second Edition, Illustrated, fcap. 8vo, price
2s. 6d.

Porteus's Evidences of the Truth and Divine
Origin of the Christian Revelation, with Definitions and Analyses by
JAMES BOYD, LL.D., late one of the Masters of the High School of
Edinburgh. 18mo, cloth, price 1s.

History of Palestine. By JOHN KITTO, D.D.
From the Patriarchal Age to the Present Time; containing Introductory Chapters on the Geography and Natural History of the Country, and on the Customs and Institutions of the Hebrews. Adapted to the purposes of Tuition by ALEX. REID, LL.D. 12mo, 3s. 6d.

STUDENTS' TEXT BOOKS.

Principles and Practice of Medicine. By J. HUGHES BENNETT, M.D., late Prof. of the Institutes of Medicine in the University of Edinburgh. Fifth Edition. 8vo, pp. 1037, illustrated with Five Hundred and Fifty Wood Engravings. Price 21s.

Lectures on Surgery. By JAMES SPENCE, Professor of Surgery in the University of Edinburgh. 2 vols. 8vo. With numerous Illustrations on Stone and in Chromo-Lithography.

Sir J. Y. SIMPSON, Bart., late Professor of Midwifery in the University of Edinburgh. In 3 vols., 8vo, price 50s.
Vol. I.—SELECT OBSTETRIC and GYNÆCOLOGICAL WORKS. Edited by J. WATT BLACK, M.D., Physician-Accoucheur in Charing-Cross Hospital, London. Demy 8vo, price 18s.
Vol. II.—ANÆSTHESIA, HOSPITALISM, ETC. Edited by Sir WALTER SIMPSON, Bart. Demy 8vo, price 14s.
Vol. III.—The DISEASES of WOMEN. Edited by ALEXANDER SIMPSON, M.D., Professor of Midwifery in the University of Edinburgh. Demy 8vo, price 18s.

Anatomical Memoirs of John Goodsir. Edited by WILLIAM TURNER, M.B., Professor of Anatomy in the University of Edinburgh. 2 vols. 8vo. Illustrated with 14 Plates, Woodcuts, and Portrait. Price 21s.

Researches in Obstetrics. By J. MATTHEWS DUNCAN, M.D., St. Bartholomew's Hospital, London. 8vo. Price 18s.

Mansel (Professor H. L.) Metaphysics, or the Philosophy of Consciousness, Phenomenal and Real, with Index. Third Edition. Crown 8vo, cloth, price 7s. 6d.

Physical Geography. By Sir J. F. W. HERSCHEL, Bart. Crown 8vo, price 5s.

An Introduction to Human Anatomy. By WM. TURNER, Professor of Anatomy in the University of Edinburgh. Crown 8vo., cloth, with 233 wood engravings. Price 16s.

Ethical Philosophy. By Sir JAMES MACKINTOSH. Edited by WILLIAM WHEWELL, D.D. New Edition (4th). Crown 8vo, price 7s. 6d.

EDINBURGH: A. AND C. BLACK.
LONDON: LONGMANS, GREEN, READER, AND DYER,
39 PATERNOSTER ROW.

www.ingramcontent.com/pod-product-compliance
Lightning Source LLC
Chambersburg PA
CBHW021947160426
43195CB00011B/1259